Schematic Wiring
A Step-By-Step Guide

Stanley H. Aglow

Business News Publishing Company
BNP Troy, Michigan

Editor: Joanna Turpin
Production Coordinator: Mark Leibold

Library of Congress Cataloging in Publication Data

Aglow, Stanley H., 1927 -
Schematic wiring, a step-by-step guide/Stanley H. Aglow
p. cm.
ISBN 0-912524-66-9 (soft bound)
1. Electric apparatus and appliances - Maintenance and repair. 2. Electric wiring - Charts, diagrams, etc. I. Title.
TK452.A33 1991
683'.83--dc20 91-39158
CIP

Printed in United States
7 6

CONTENTS

Section I

INTRODUCTION

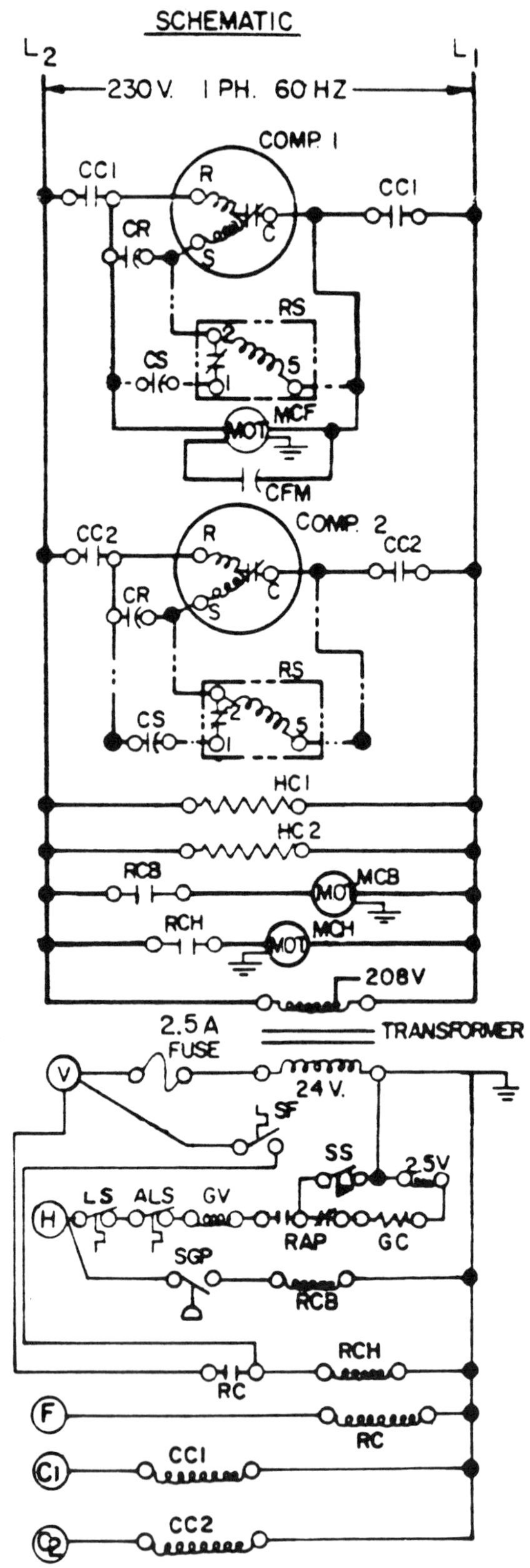

Figure 1-1. Typical schematic diagram. Courtesy, *Fedders Corporation.*

ELECTRICAL SCHEMATICS

The schematic, Figure 1-1, is the engineer's blueprint of an electrical circuit. It differs from a builder's blueprint in that it does not show the actual location of the electrical devices used in the circuit, but only the electrical relationship of one device to another.

Figure 1-2a is a pictorial drawing of a light bulb, with a cord, plug and switch. Figure 1-2b is an electrical schematic of the same light bulb, cord, plug and switch. Instead of a plug, this drawing uses L_1 and N as the power source and a single-pole, single-throw (SPST) switch to control the current flow in one conductor to turn the light on and off. From the schematic, you can see how the switch is wired into the circuit, while the pictorial representation shows only the location of each device in the circuit. You can see that this additional information makes the schematic invaluable when troubleshooting a circuit.

The schematic, which is sometimes called a ladder drawing, uses symbols and abbreviations in an unusual type of shorthand. The meaning of these symbols and abbreviations is usually spelled out in a feature of the schematic called the legend.

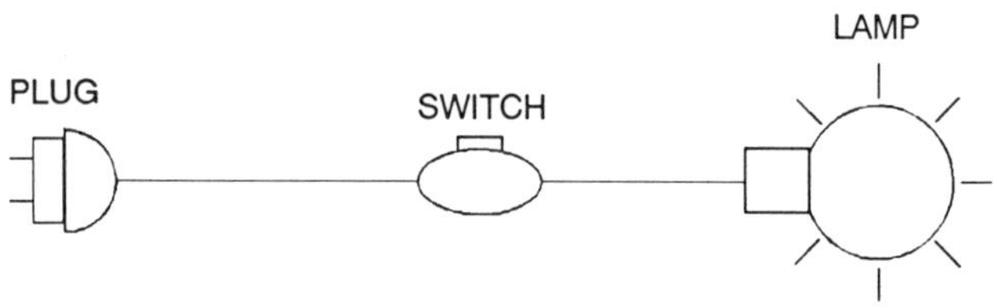

Figure 1-2a. Pictorial drawing of a lamp circuit.

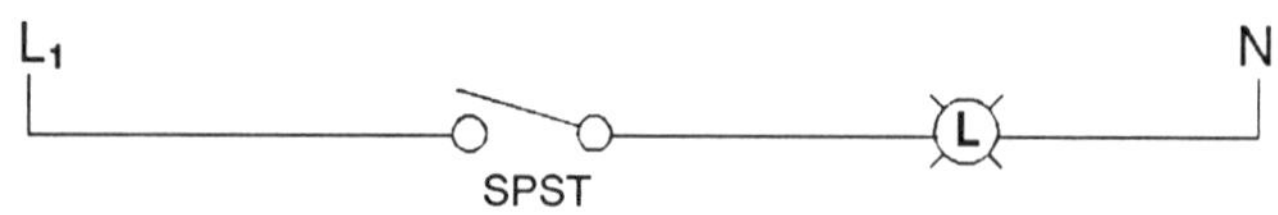

Figure 1-2b. Schematic equivalent of the pictorial drawing in Figure 1-2a.

THE LEGEND

The legend, or key, should list all the devices and abbreviations appearing in the schematic and give a concise description or function of any special controls. The legend may also include timing settings and other necessary information. Figure 1-3a shows the legend that accompanies the Figure 1-1 schematic. Figure 1-3b is an example of another legend that accompanies a General Electric heat pump.

You will find that there are a variety of abbreviations for the same device. For example, Carrier abbreviates an indoor fan relay as 1FR, while York uses R and Westinghouse uses FR.

You may also find a variety of designations for the same device. Carrier uses both an interchangeable holding relay (HR) and a reversing relay (RVR). Both relays have 208/240-V coils and two sets of normally-open and normally-closed contacts. They are essentially the same device; however, they are used for different purposes.

In addition to the legend, you may find notes on the schematic. These notes include exceptions, changes in wiring for varying voltages, motor protection devices being used, and wires that must be altered when changing motor speeds. This information is valuable when troubleshooting or installing the equipment.

An important part of the legend, especially for the installer, is the wiring identification which is basically the same for all manufacturers:

- Heavy black line (High voltage, factory wired)
- Light black line (Low voltage, factory wired)
- Heavy broken line (High voltage, field wired)
- Light broken line (Low voltage, field wired)

CC - CONTACTOR COMPRESSOR
COMP - COMPRESSOR
CFM - CAPACITOR FAN MOTOR
CR - CAPACITOR RUN (COMP.)
CS - CAPACITOR START (COMP.)
GV - GAS VALVE
HC - HEATER COMPRESSOR
LS - LIMIT SWITCH
MCB - MOTOR COMBUSTION BLOWER
MCF - MOTOR CONDENSER FAN
MCH - MOTOR HEATING/COOLING BLOWER
RAP - RELAY AUTO PILOT
RCB - RELAY COMBUSTION BLOWER
RCH - RELAY COOLING/HEATING BLOWER
RS - RELAY START
SF - SWITCH FAN
SGP - SWITCH PILOT GAS PRESSURE
SS - SWITCH SAIL
GC - GLOW COIL
ASL - AUXILIARY LIMIT SWITCH
RC - RELAY COOLING

Figure 1-3a. Legend accompanying the Figure 1-1. schematic. Courtesy, *Fedders Corporation.*

CA - COOLING ANTICIPATOR
CF - FAN CAPACITOR
CN - WIRE CONNECTOR
CR - RUN CAPACITOR
CPR - COMPRESSOR
CS - START CAPACITOR
CSR - CAPACITOR SWITCHING RELAY
D - DEFROST RELAY
DFT - DEFROST TIMER
DT - DEFROST TERMINATION SWITCH
F - INDOOR FAN RELAY
FM - MANUAL FAN SWITCH
FU - FUSE
HA - HEATING ANTICIPATOR
IOL - INTERNAL OVERLOAD PROTECTOR
LT - LIGHT
MS - COMPRESSOR MOTOR CONTACTOR
ODA - OUTDOOR TEMPERATURE ANTICIPATOR
ODS - OUTDOOR TEMPERATURE SENSOR
ODT - OUTDOOR THERMOSTAT
RHS - RESISTANCE HEAT SWITCH
SC - SWITCHOVER VALVE SOLENOID
SM - SYSTEM "ON-OFF" SWITCH
TNS - TRANSFORMER
TS - HEATING & COOLING THERMOSTAT
TSC - COOLING THERMOSTAT
TSH - HEATING THERMOSTAT
UA - UNIVERSAL ANTICIPATOR (HEATING & COOLING)

Figure 1-3b. Legend accompanying Heat Pump. Courtesy, *General Electric Company.*

ELECTRICAL SYMBOLS

The schematic makes use of many standard electrical symbols. Each manufacturer may prefer slight variations, but they are all basically the same.

The combination of the graphic symbols and the abbreviations found in the legend form an electrical shorthand which, when mastered, allows you to troubleshoot electrical circuits much faster and with fewer errors. Some of these symbols are shown in Figures 1-4 and 1-5. Notice that the name of each device is shown with the symbol.

Contacts in a control are shown in their normal positions. The normal position assumes that the system is off or de-energized. Thus, the normal position of an electromagnetic relay, when its coil is de-energized, is with the contacts open.

The normal position for a pressure control assumes that it is affected by atmospheric pressure only. When the pressure control is installed, it opens or closes as the pressure in the system changes.

Flow switches, whether gas or liquid, are in the normal position when no liquid is moving. Movement of the liquid causes the contacts to change position.

Figure 1-6 shows that the position of the moveable element in the contact symbol has a specific meaning:

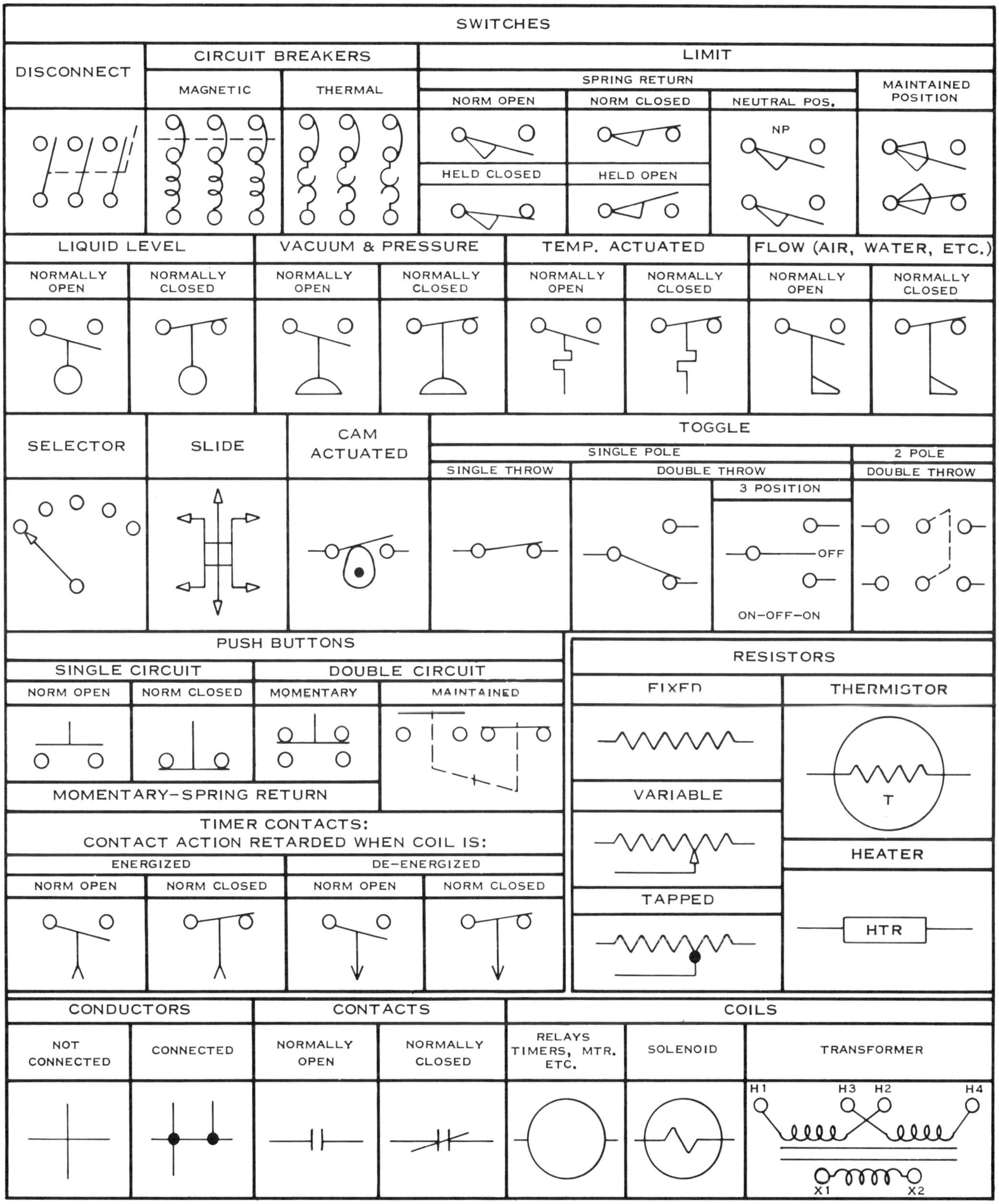

Figure 1-4. Typical graphic symbols for electrical schematics. Courtesy, *York Division, Borg-Warner Corporation.*

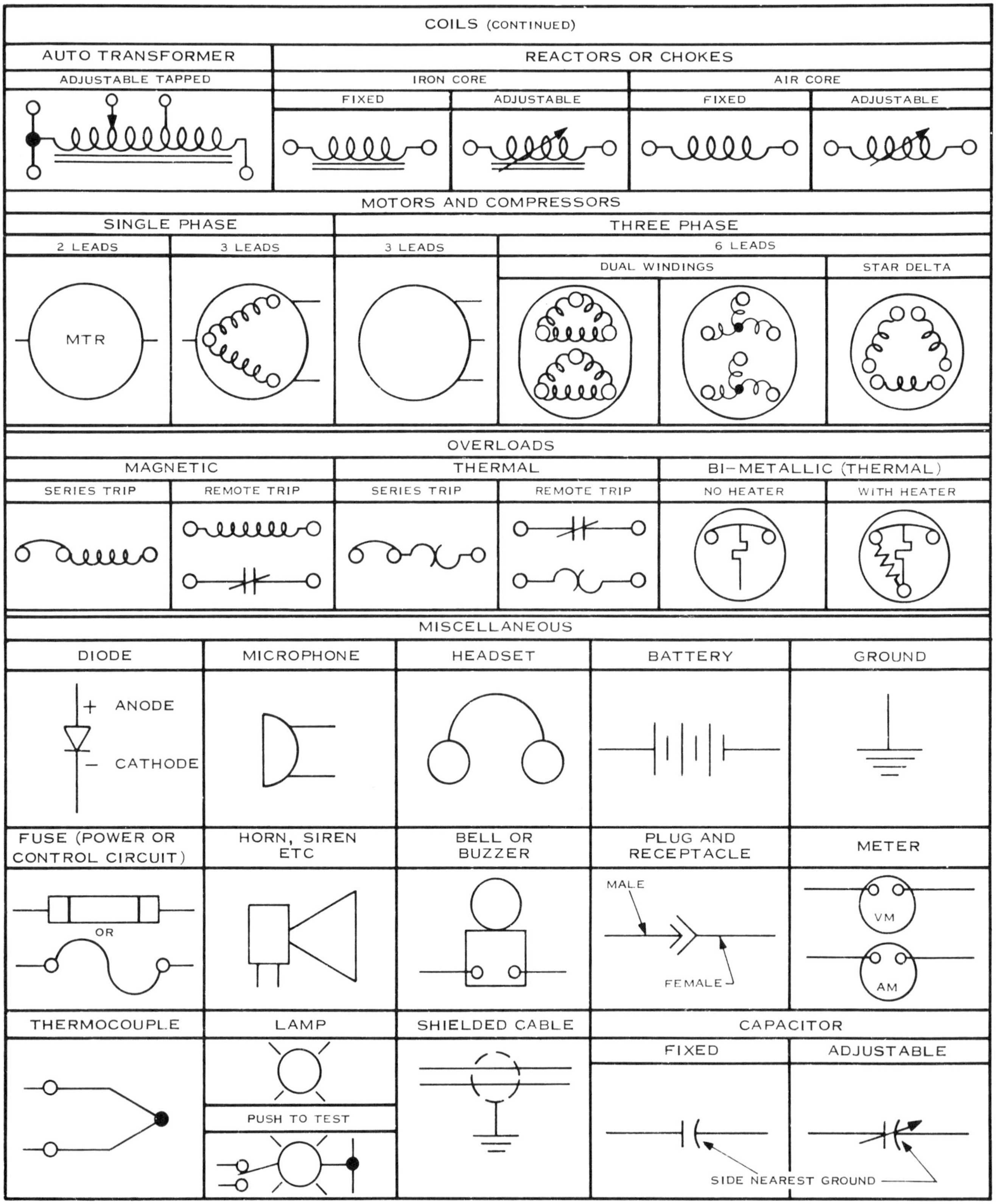

Figure 1-5. Typical graphic symbols for electrical schematics. Courtesy, *York Division, Borg- Warner Corporation.*

1. A moveable contact that is open and below the fixed contact indicates that the switch closes on an increase in temperature, pressure, etc.
2. A moveable contact that is closed and below the fixed contact indicates that the switch opens on a decrease in temperature, pressure, etc.
3. A moveable contact that is open and above the fixed contact indicates that the switch closes on a decrease in temperature, pressure, etc.
4. A moveable contact that is closed and above the fixed contact indicates that the switch opens on an increase in temperature, pressure, etc.

You should spend as much time as necessary to commit these symbols to memory. Some manufacturers of heating, cooling and refrigeration equipment do not show the symbols in their service manuals and most mechanics do not have a copy of the symbols in their tool box. Without understanding the symbols, troubleshooting becomes impossible.

SYMBOL	NAME	CONTACT POSITION	FORCE
	LOW PRESSURE SWITCH	NO	CLOSES ON INCREASE IN REFRIGERANT PRESSURE
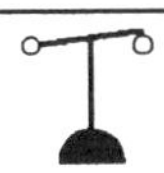	HIGH PRESSURE SWITCH	NC	OPENS ON INCREASE IN REFRIGERANT PRESSURE
	COOLING THERMOSTAT	NO	CLOSES ON RISE IN TEMPERATURE
	HEATING THERMOSTAT	NC	OPENS ON RISE IN TEMPERATURE
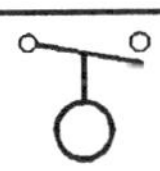	LIQUID LEVEL SWITCH	NO	CLOSES ON RISE IN LIQUID LEVEL
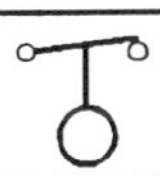	LIQUID LEVEL SWITCH	NC	OPENS ON RISE IN LIQUID LEVEL
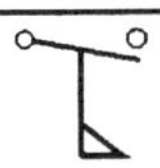	FLOW SWITCH LIQUID OR AIR	NO	CLOSES ON FLOW OR MOVEMENT
	FLOW SWITCH LIQUID OR AIR	NC	OPENS ON FLOW OR MOVEMENT

Figure 1-6. Interpreting symbols for moveable contacts.

ELECTRONIC SYMBOLS

The use of electronic controls in the HVAC-R industry is increasing rapidly. The high cost of energy and the industry's desire to produce more energy-efficient equipment has led to this great increase. Solid state electronic safety and cycling controls have led the way to a more economical use of energy for HVAC-R. The greatest part of the electronic schematic involves the use of specialized symbols to depict electronic devices and controls, as shown in Figures 1-7 and 1-8. The HVAC-R technician must be an expert at understanding electronic and electrical schematics. The symbols pictured in this section are for electronic drawings. Some electronic symbols are similar to the electrical symbols.

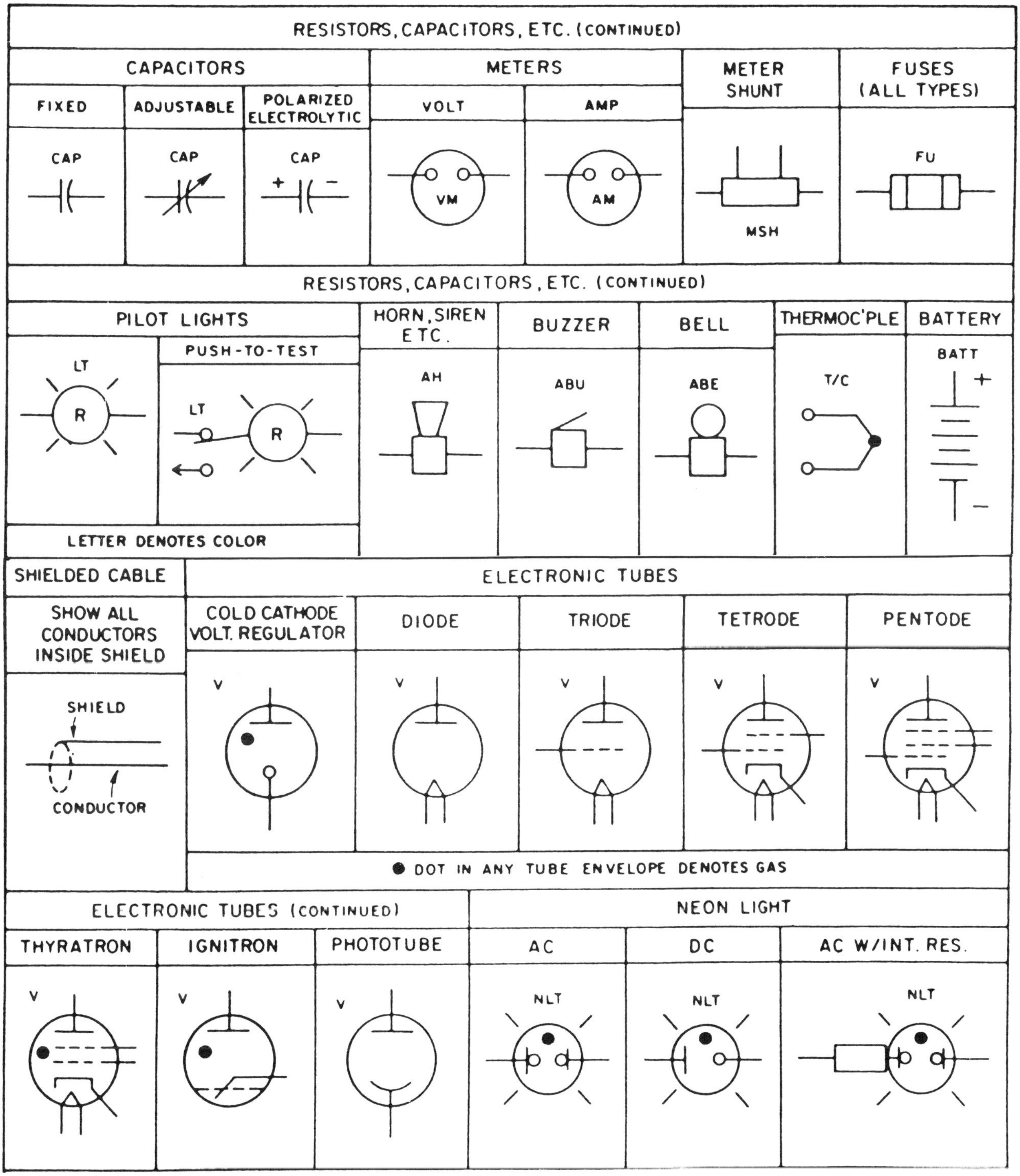

Figure 1-7. Typical graphic symbols with basic device designations. Courtesy, *National Machine Tool Builders Association.*

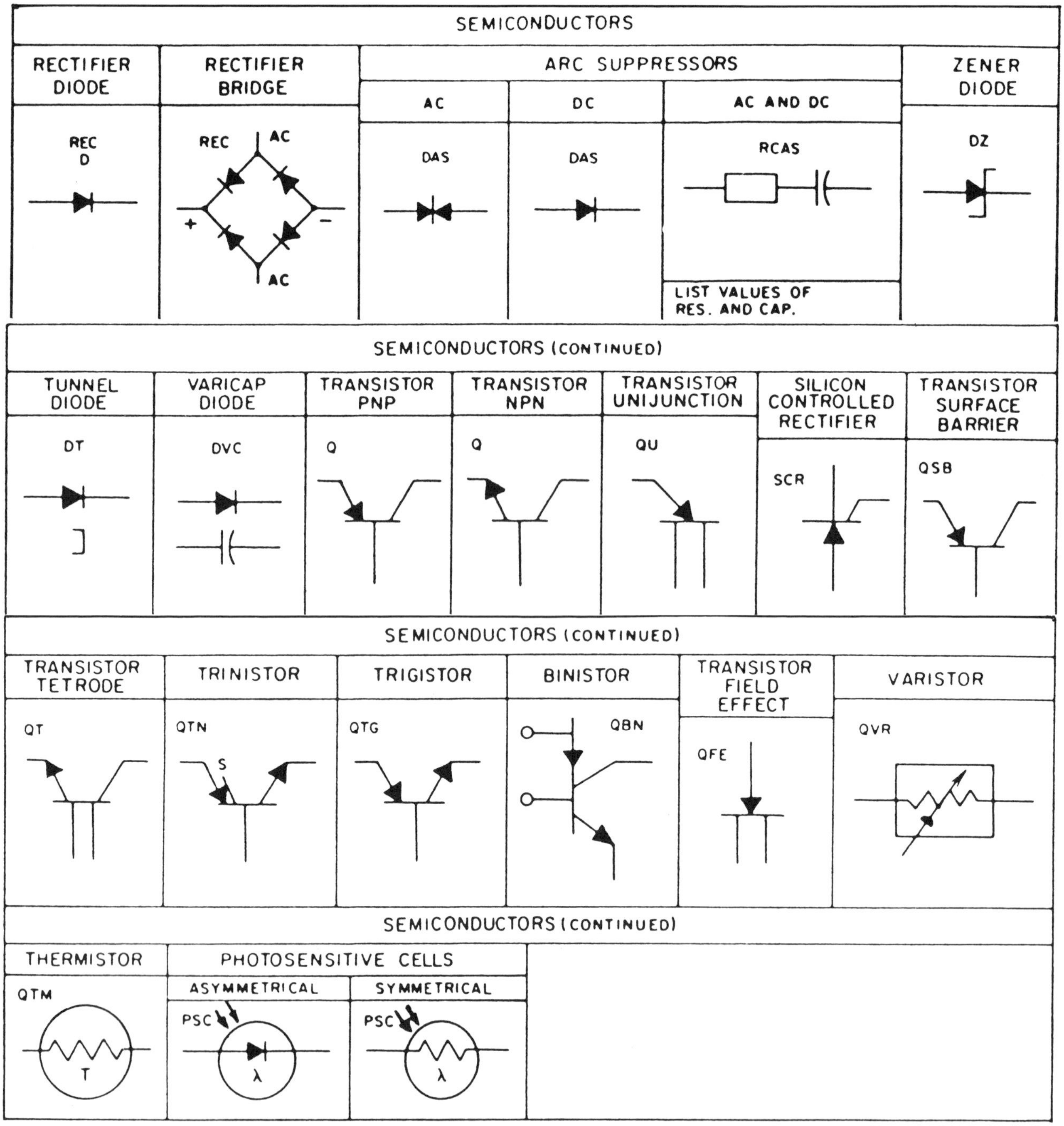

Figure 1-8. Typical graphic symbols with basic device designations. Courtesy, *National Machine Tool Builders Association.*

Section II

BASIC ELECTRICITY AND DEVICES

This section is not intended to be an in-depth study of each electrical device. Rather it serves to explain and show how each device will function in an electrical circuit.

MAGNETISM AND MAGNETS

Magnetism is the operating force behind all electromagnetic relays, motors and solenoids. Magnetism can be either permanently available or electrically produced.

Permanent magnets are created by exposing a bar of hardened steel to a strong magnetic field. The force of this field permanently aligns the molecules in the bar, giving them magnetic polarity. When the strong magnetic force is removed, the bar will continue to retain its magnetism.

Permanent magnets are commonly used as cabinet and refrigerator latches. Other permanent magnets are used in thermostats and switches to ensure snap action. The permanent magnet rarely becomes a source of trouble.

An electromagnet is created by wrapping a coil of wire around a soft iron core, Figure 2-1. When current is passed through the coil of wire, the soft iron core becomes strongly magnetized. If the current ceases, the iron core instantly loses its magnetism. The ability to quickly magnetize and demagnetize the iron core permits remote operation of many of the controls used in heating, cooling and refrigeration.

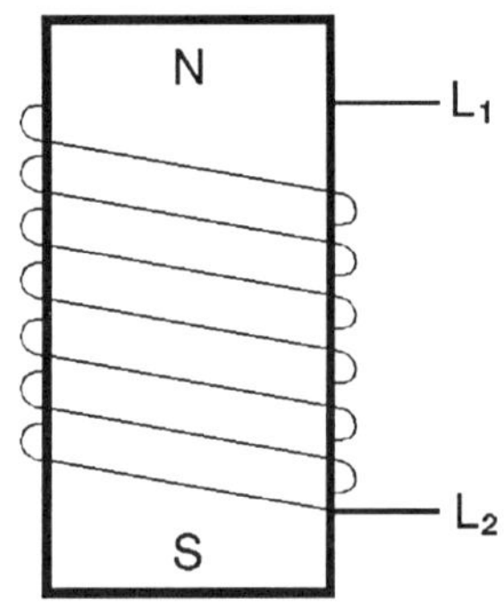

Figure 2-1. An electromagnet is created by wrapping a coil of wire around a soft iron core.

Solenoids

Most electromagnets, such as those used in relays, have a fixed coil and a fixed core. The solenoid, on the other hand, is an electromagnet with a fixed coil and a moveable core, Figure 2-2.

In one typical application, Figure 2-3, the solenoid is used to open and close a gas valve. When current is supplied to the coil, the iron core is electromagnetically drawn into the center of the coil, opening the valve. When the current is interrupted, the magnetic field is lost and the core drops, usually by gravity, to again close the valve.

Solenoid valves operate on voltages ranging from less than one volt to over 440 V, in a range of applications, Figure 2-4.

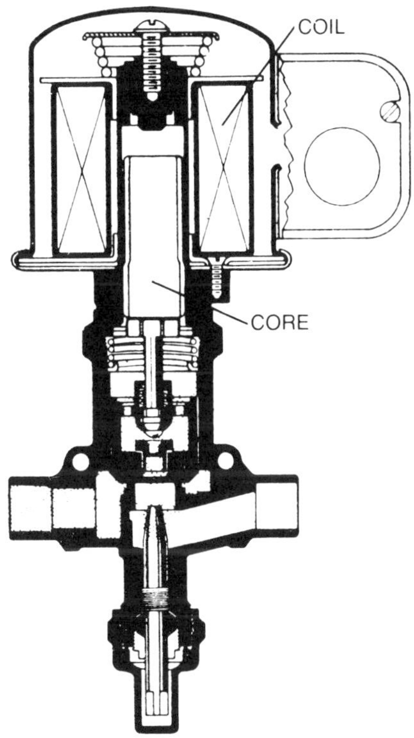

Figure 2-3. A solenoid operated gas valve. Courtesy, *ITT General Controls.*

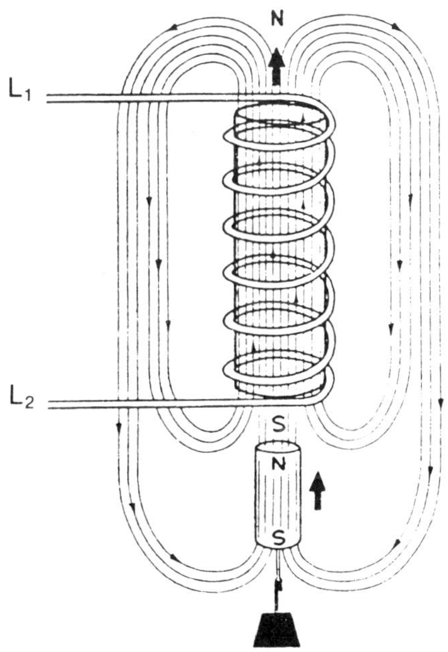

Figure 2-2. A solenoid is an electromagnet with a moveable soft iron core.

Figure 2-4. A solenoid used for refrigerant flow control. Courtesy, *Alco Controls Division, Emerson Electric Company.*

TRANSFORMERS

In refrigeration, cooling, and heating applications, it is often necessary to use voltages lower or higher than those supplied by the electric company. It is necessary, then, to modify this power supply to create the required voltage. The least expensive way to do this is with a transformer, Figure 2-5a.

The transformer has two unconnected windings, a primary and a secondary, each of which is usually wound around a core of magnetic material. An alternating current (ac), when applied to the primary, induces voltage in the secondary winding.

A secondary that has fewer turns than the primary is called a step-down transformer. The step-down transformer is used for low voltage thermostats and control circuits. A transformer having more turns in the secondary than in the primary is called a step-up transformer, Figure 2-5b. Step-up transformers are used when a higher voltage is needed, for example, 10,000 V.

Due to coil resistance losses in the transformer, the voltage times amps (VA) of the primary and secondary coils are unequal, causing the transformer to be less than 100% effective. These losses must be considered when sizing a transformer. The secondary needs to have enough electrical energy to drive the connected load. System equipment and amperage ratings must be considered when selecting the proper transformer.

Consider the secondary of the transformer to be the same as the service entrance cable to your home. The size of this cable is based on the amperage draw (60A, 100A, 150A, 200A, etc.) at the voltage (120 V, 208 V, 240 V, etc.) to operate all of the electrical equipment in your home.

Figure 2-5a. Step-down transformer with mounting plate. Courtesy, *Honeywell Incorporated.*

Figure 2-5b. Step-up transformer. Courtesy, *Allanson, Incorporated.*

The transformer VA can be calculated by adding the amperage rating on each magnetic relay coil, solenoid, etc. This gives you the total current draw of the entire system. Multiply this total by the voltage necessary, such as 24 V, gives you the total VA required.

A transformer must meet two requirements:

1. It must be capable of pulling in all magnetic devices at once, under inrush or starting conditions.
2. It must be capable of carrying the continuous current draw of all components, without overheating.

The transformer is the same as the power supply. For this reason, the wire used in the secondary or low voltage circuit must be properly sized to prevent a voltage drop at the loads.

There is no physical connection between the primary and secondary coils on a transformer, so the coils can be checked with an ohmmeter. This determines whether there are open or shorted windings. A transformer's output voltage that is within 10% of its rating is considered good.

Transformers that operate on more than one primary voltage, such as 120 V, 208 V, or 240 V, are called tapped transformers. Various primary voltages may be applied by tapping the transformer primary at selected points within its length, Figure 2-6. Each tap produces 24 V, because the ratio of the number of turns in the primary and secondary coils determines the secondary output. Figure 2-7 illustrates one of these transformers.

$$\frac{\text{secondary turns}}{\text{primary turns}} \times \text{primary voltage} = \text{secondary volts}$$

$$\frac{100}{1000} \times 240V = 24V$$

$$\frac{100}{850} \times 208V = 24V$$

$$\frac{100}{500} \times 320V = 24V$$

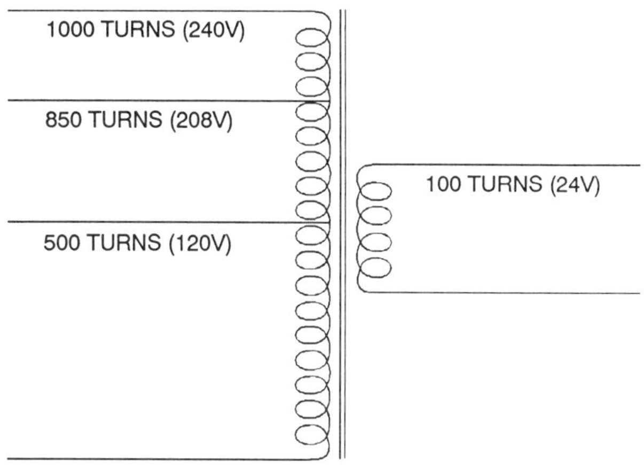

Figure 2-6. A tapped transformer that produces 24v from a variety of primary voltages.

Figure 2-7. Transformer with three primaries and one secondary. Courtesy, *Honeywell, Incorporated.*

RELAYS

The electromagnetic relay, Figure 2-8, is a magnetically-operated switch. A typical relay consists of a low (24 V) voltage electromagnet and one or more sets of high (115/230 V) voltage contacts.

The contacts may be normally open (NO) or normally closed (NC), or a combination of the two. When the electromagnet is energized, all the contacts in the relay are reversed: that is, all NC contacts open and all NO contacts close. When the relay is de-energized, all contacts return to their normal positions, whether open or closed.

Figure 2-8 shows that the relay coil and contacts are both mounted on a common base. If a wiring diagram were included here, it would show both the coil and contacts in the same location. However, a schematic diagram would separate the coil from the contacts to show the coil in the low voltage circuit and the contacts in the high voltage circuit.

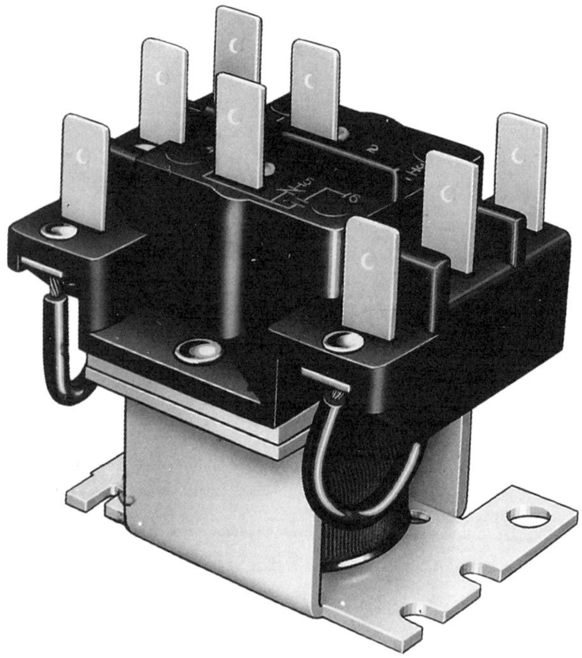

Figure 2-8. Electromagnetic relay with two sets of NO and two sets of NC contacts. Courtesy, *Essex Controls.*

Contactor and Motor Starter

The contactor, Figure 2-9, which is basically a relay, makes and breaks the power supply to the compressor. If the contact points are rated high enough, the condenser fan motor may also be wired through the contactor, parallel to the compressor. This configuration allows both the condenser fan and the compressor to start at the same time. Contactors switch heavy current loads, high voltage or both. Because the coil is isolated from the contacts, a low voltage (24 V or 120 V) can be supplied to the coil, and 208 V, 240 V, 277 V, or 480 V can be supplied through the contacts to the load.

A motor starter is basically a contactor with the addition of overload heaters.

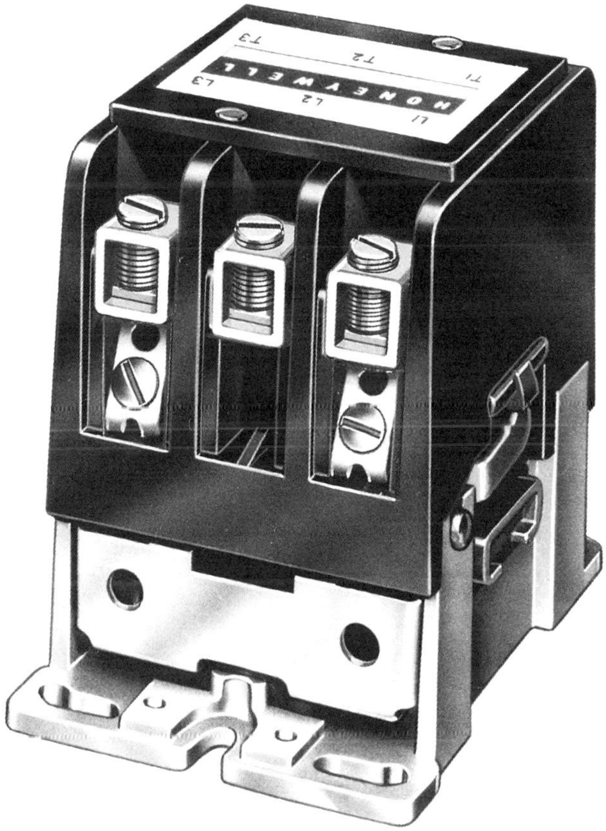

Figure 2-9. A contactor for switching on motors or resistive loads. Courtesy, *Honeywell Incorporated.*

POTENTIAL MAGNETIC STARTING RELAY

The potential or voltage starting relay, Figure 2-10a, is used on single-phase, capacitor start, hermetic compressors. This relay closes a circuit to the phase or starting winding, until the motor reaches approximately 80% of its rated speed. It then opens the circuit to the phase winding. Figure 2-10b is the schematic wiring diagram of Figure 2-10a.

The potential relay consists of a fine wire electromagnetic coil, with a voltage rating above line voltage. The contact points are normally closed, thus completing a circuit through the start capacitor, from the run to start terminals on the compressor.

When the motor starts, the counter-electromagnetic force (emf), or back voltage, quickly increases in the starting winding. This causes an increase in voltage through the relay coil, which is wired between the common and start terminals on the compressor. When the rated voltage of the relay coil is reached, the coil becomes an electromagnet, opening the normally-closed contacts.

During the running cycle, the back voltage remains high enough to hold the contacts open. If the motor slows down because of a heavy load, a defective run capacitor, or an open motor circuit, the voltage feeding the potential relay coil will drop rapidly. This de-energizes the coil, permitting the relay contacts to close.

Figure 2-10a. A potential magnetic starting relay. Courtesy, *Essex Controls.*

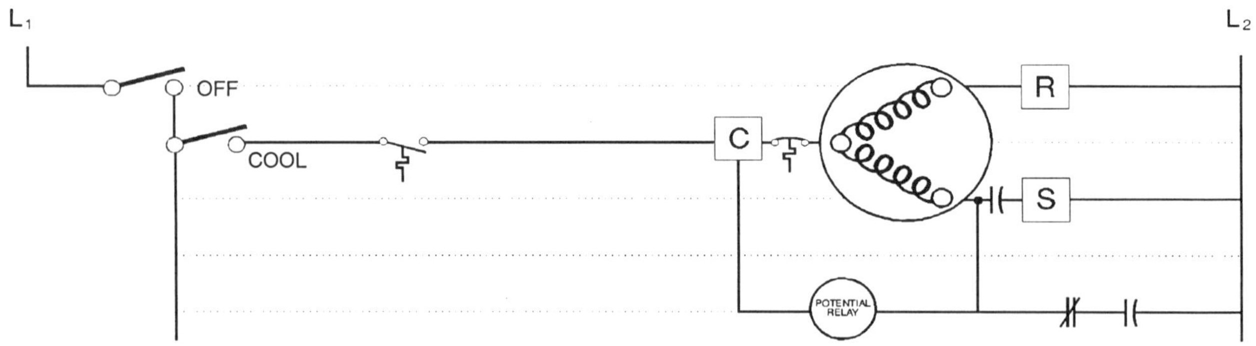

Figure 2-10b. Schematic wiring for potential relay.

Current Magnetic Starting Relay

The current or amperage starting relay, Figure 2-11a, is used on fractional horsepower (hp), single- phase, hermetic compressors. The relay closes a circuit to the phase or starting winding until the motor reaches approximately 2/3 or 3/4 of its rated speed. The relay then opens the circuit to the starting winding. Figure 2-11b is a schematic wiring diagram of Figure 2-11a.

Current relays consist of a heavy wire coil and a set of normally-open contacts. When a circuit to the motor is closed, all current is fed to the main or running winding, passing through the start relay coil. The running winding, alone, cannot start the motor, which creates a locked rotor condition. This causes high amperage or current to flow through the motor winding and the relay coil, producing a strong magnetic field in the relay coil. This condition then closes the normally-open contacts, which completes the circuit to the starting winding and starting the motor.

As the motor approaches its rated speed, the current draw of the motor drops, and the coil's magnetic field becomes too weak to hold the contacts closed. The contacts then open, breaking the circuit to the start winding. The relay contacts stay closed for only a few seconds each time the motor starts.

Figure 2-11a. Current magnetic starting relay. Courtesy, *Texas Instruments, Incorporated.*

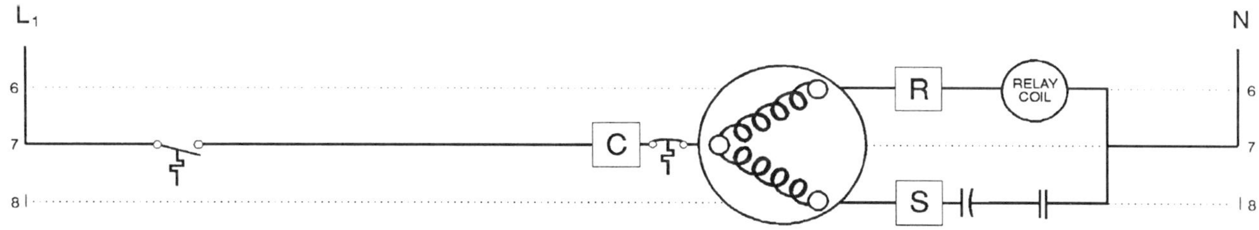

Figure 2-11b. Schematic wiring for current. Magnetic starting relay.

INDOOR FAN RELAY

The indoor fan relay consists of an electromagnetic coil and a set of normally-open and normally-closed contacts.

The fan relay starts the indoor or evaporator fan whenever the cooling compressor starts. If the evaporator blower does not start, the evaporator can freeze up and possibly ruin the compressor. A fan relay must be installed whenever an air conditioning system is added to an existing furnace.

The normally-open contacts of the fan relay, Figure 2-12, must be wired parallel to the plenum switch. The normally-closed contacts must be wired in series with the plenum switch.

When the thermostat calls for heat, the temperature-activated plenum switch, which closes at approximately 140 °F, cycles the fan. After the plenum switch is closed, normally-closed contacts supply power to the fan.

When cooling is called for, 24-V power is supplied to the indoor fan relay coil. Energizing the coil causes the normally-closed contacts to open and the normally-open contacts to close. The power bypasses the plenum switch and activates the fan, which is then controlled by the cooling thermostat.

A fan relay can be used to select the desired speed from a two-speed motor. As shown in Figure 2-13, the fan relay selects low speed while heating and high speed while cooling. The relay prevents both speeds from being energized simultaneously, which could destroy the motor.

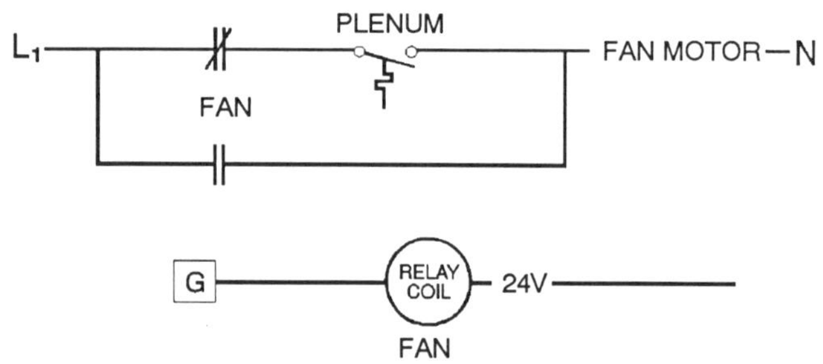

Figure 2-12. Schematic of an indoor fan relay.

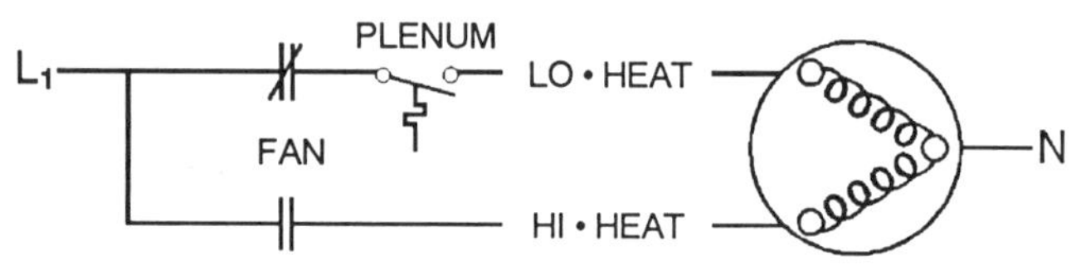

Figure 2-13. Indoor fan relay can select the desired speed from a two-speed motor.

FAN CENTER

The fan center is a relay with a built-in, 24-V step-down transformer, Figure 2-14. Figure 2-15 shows an actual fan center.

When installing a fan center, connect the primary of the transformer to the 120-V, or higher line, power supply (L_1 and N). Connect the secondary, or 24-V, side of the transformer to thermostat terminal R, or common. Connect the fan center relay coil to thermostat terminal G, which is the fan terminal. Connect the normally-closed contacts in series with the furnace plenum switch and the normally-open contacts to the fan motor. Both sets of contacts are in the line voltage circuit.

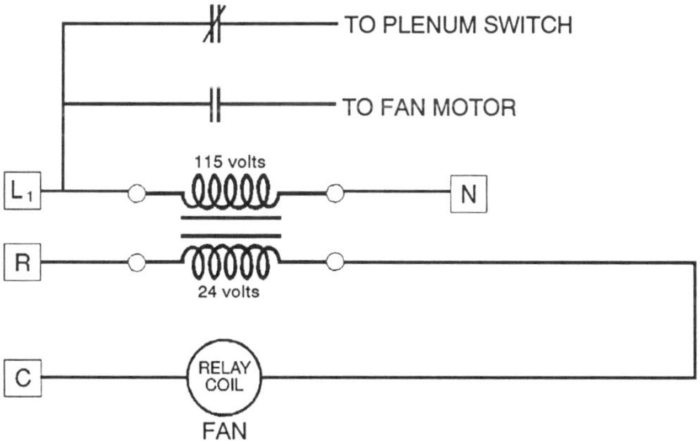

Figure 2-14. Schematic of a fan center.

Figure 2-15. Fan center, Mod 90-113 Type 631. Courtesy, *White Rodgers Division.*

IMPEDANCE OR LOCKOUT RELAY

Impedance relays, Figure 2-16, provide a lockout circuit and remote manual reset in refrigeration and air conditioning systems. When a safety control, such as a high or low pressure switch, motor overload, oil safety switch, etc., opens, the lockout circuit prevents the system from starting, even after all safety control contacts close. Locking out the control circuit allows you to determine the problem before more damage is done.

The relay consists of a set of normally-closed contacts wired in series with the safety controls and the compressor contactor coil, Figure 2-17. The impedance coil is wired in series with the contactor coil and parallel to the relay's normally- closed contacts and the closed contacts of the safety controls. During normal operation, the normally-closed contacts of the relay and the safety controls present a course of least resistance, permitting full voltage to energize the contactor coil.

If a pressure control or overload opens, the shunt circuit to the contactor is broken, energizing the impedance coil. The resistance of the impedance coil is so high that there is not enough voltage to the contactor coil to hold in the contactor contacts. As the impedance relay pulls in, its normally-closed contacts open, locking out the circuit to the contactor coil, even though the safety controls may have reset.

The system will remain locked out and can only be started again by breaking the circuit to the contactor coil. This is usually done by moving the thermostat control switch to the off or reset position and then back to cool.

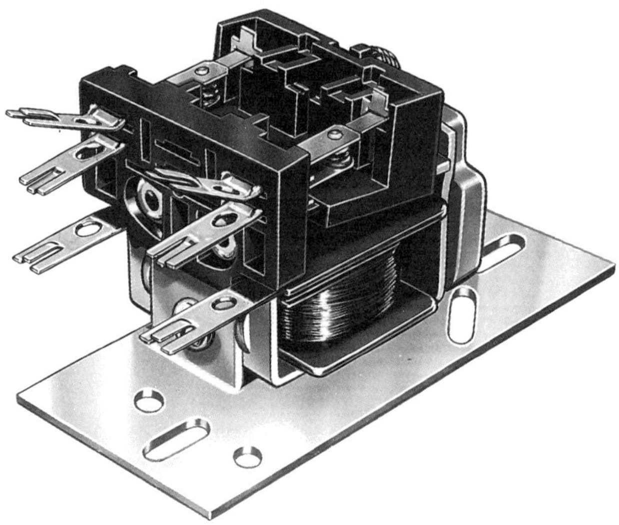

Figure 2-16. Impedance or lockout relay. Courtesy, *Honeywell Incorporated.*

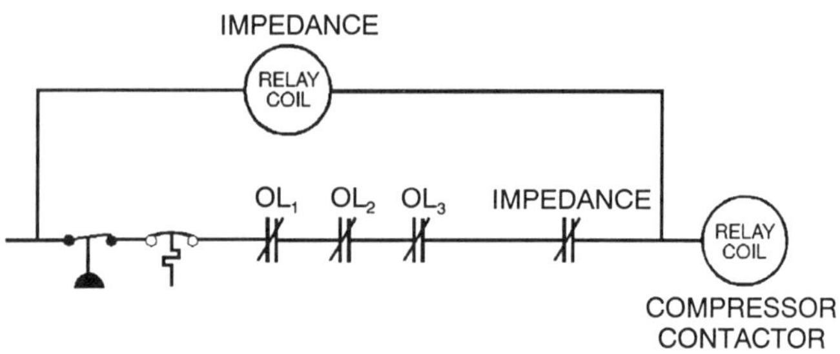

Figure 2-17. In an impedance relay circuit, the impedance coil is wired in series with compressor contactor coil and parallel to the relay's normally-closed contacts.

THERMAL TIME DELAY RELAY

The thermal time delay relay, Figure 2-18, consists of a heater coil wound around a bimetal element and a set of normally-open or normally-closed contacts, Figure 2-19. The heater element comes in many different voltages, such as 24 V, 120 V, 208 V, 240 V, etc. Because there is no internal electrical connection between the heater and the contacts, the voltage used for the heater circuit (24 V) may differ from the voltage going through the contacts (240 V).

The heater warms up after a time delay, which causes the bimetal element to bend. This action opens or closes the contacts, depending on the application, with a snap action. Snap action is accomplished by using a small permanent magnet. This action prevents chattering or arcing of the contact points on make and break.

Time delay relays are used to prevent two or more compressors in one system from starting at the same moment. This eliminates the excessive inrush of current and ensures that the water pump starts before the compressor in a water-cooled system, Figure 2-20.

Time delay relays are also used on electric resistance furnaces or heaters. This allows each heat strip to start in sequence, preventing an excessive inrush of current, which would occur if all the strips were energized at once.

A 4 to 5 minute time delay relay prevents short cycling of a hermetic compressor after it shuts down due to the safety controls or thermostat. This means the relay will not permit the compressor to start for 4 to 5 minutes, reducing the chances of overheating the hermetic compressor motor and causing a burnout. Time delay relays are also used to control the incremental starting in a large, three-phase motor.

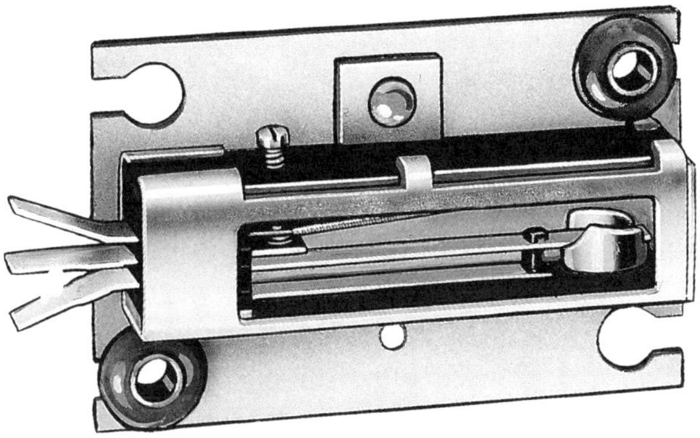

Figure 2-18. Cutaway of a thermal time delay heater. Courtesy, *Essex Controls.*

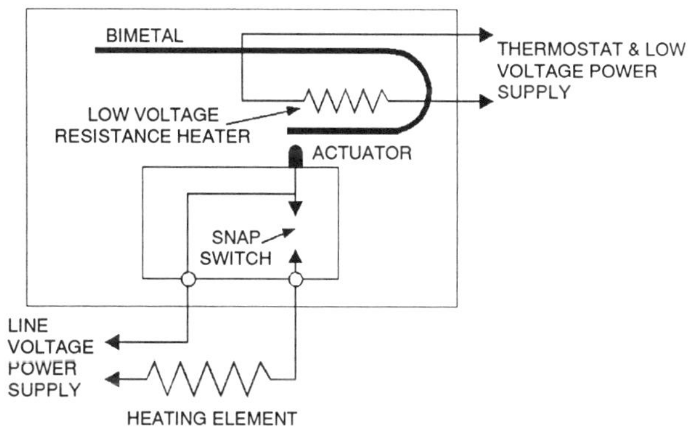

Figure 2-19. Schematic of a thermal time delay heater. Courtesy, *Honeywell Incorporated.*

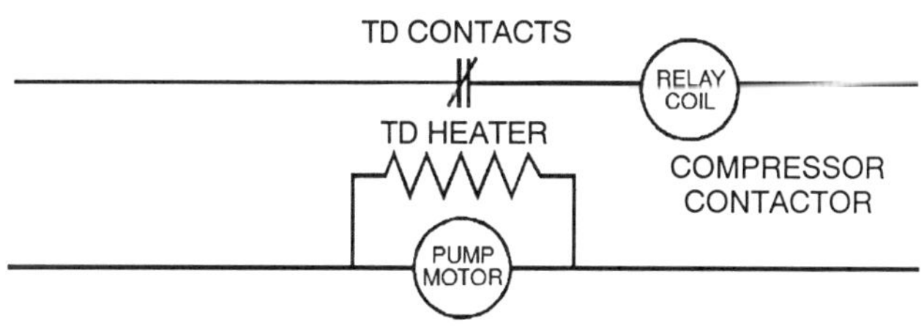

Figure 2-20. When the water pump contacts (not shown) close, both the water pump and thermal time delay (TD) heater are energized. 60 seconds later, the heater will warp closed the TD contacts and complete a circuit to the compressor contactor relay. In this example, line voltage is supplied to both the TD heater and the TD contacts.

THERMOSTATS

The thermostat is usually the main controller of the heating or cooling system. It senses the temperature of the surrounding air and opens or closes a circuit to start or stop the heating or cooling equipment. This maintains the thermostat temperature setting.

Thermostats may be either line or low voltage, but low voltage thermostats are typically used in residential applications. This is because the low voltage thermostat has better control characteristics, is easier and less expensive to install, and is safer to use in the home.

The bimetal element, is the most common device used to detect thermal change. The bimetal consists of two thin strips of dissimilar metals that are bonded together. One metal has a higher rate of expansion than the other, so when exposed to heat, the two metals expand at different rates. This expanding causes the bimetal to bend or warp. When one end of the strip is anchored, the other end will bend enough to open or close a set of contacts as the temperature changes.

These contacts, when attached to the bimetal strip, may carry the electrical load. Or, a small permanent magnet may be used to assure snap action by a separate set of contacts.

The most popular method of making and breaking the electrical circuit in a low-voltage thermostat is to attach a mercury bulb to the moveable end of the bimetal element, Figure 2-21. As the bimetal moves, the mercury in the bulb flows to one side or another. This action causes the mercury to open or close a set of contacts, which then makes or breaks the electrical circuit. Unlike the straight bimetal that can be mounted in any position, the mercury bulb must be mounted upright and level for proper performance.

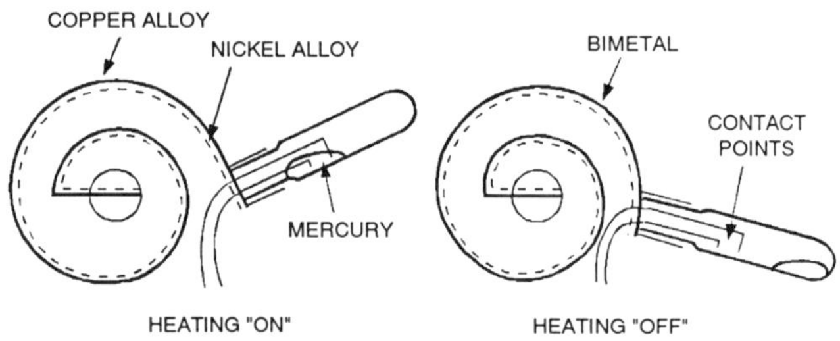

Figure 2-21. When the bimetal contracts, on a fall in temperature, the mercury in the bulb bridges the two contacts to complete the heating circuit. Courtesy, *Honeywell Incorporated.*

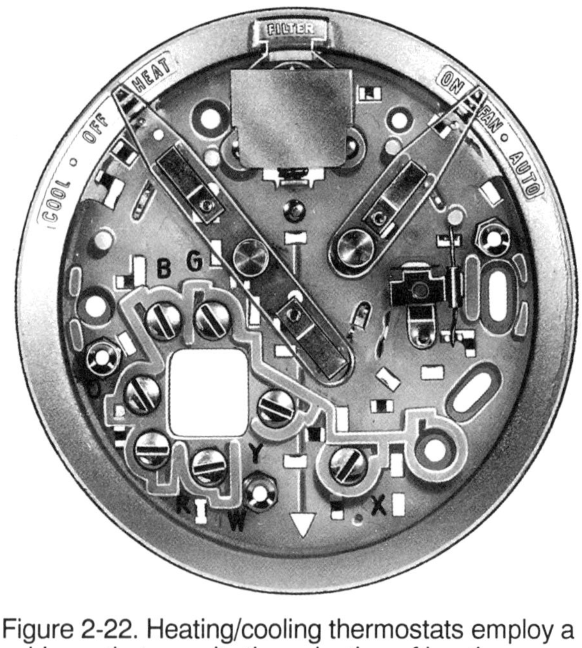

Figure 2-22. Heating/cooling thermostats employ a subbase that permits the selection of heating or cooling and automatic or constant fan operation. Courtesy, *Honeywell Incorporated.*

The heating/cooling thermostat usually comes with a subbase, Figure 2-22. This subbase comes equipped with HEAT, OFF, and COOL positions, as well as two positions for fan operation - ON and AUTO. When the thermostat is in the COOL position, the indoor or evaporator fan must also be on, to prevent the evaporator from frosting. When in the HEAT position, the indoor fan must start when the furnace plenum temperature reaches 140 °F. The AUTO fan position operates only when the heating or cooling system is on. In the On position, the fan runs constantly, even when the system is in the standby mode.

To provide a more comfortable room temperature, thermostats usually have heating and cooling anticipation, Figure 2-23. Heating anticipation is accomplished by wiring a small variable heating resistor in series with the heating contacts in the thermostat, Figure 2-24. The anticipator, set to match the amperage rating of the primary control, adds heat to the bimetal element when the heating contacts close. This added heat, plus the room temperature, causes the heating contacts to open before the room comes up to the thermostat setting. The residual heat in the heater brings the room up to the thermostat setting, preventing overheating.

The cooling anticipator is a fixed resistor, wired parallel to the cooling contacts, Figure 2-25. This anticipator adds heat to the bimetal element when the cooling contacts are open. The added heat, plus the rising room temperature, causes the cooling contacts to close a little before the actual room temperature reaches the thermostat setting. When the cooling contacts close, the cooling anticipator is bypassed, and the cooling contacts open when the air temperature drops to the thermostat setting.

Because the heating anticipator is wired in series with the heating contacts, a break in the anticipator element causes a shutdown of the heating system. A break in the cooling anticipator, wired parallel to the cooling contacts, does not cause a shutdown of the cooling system.

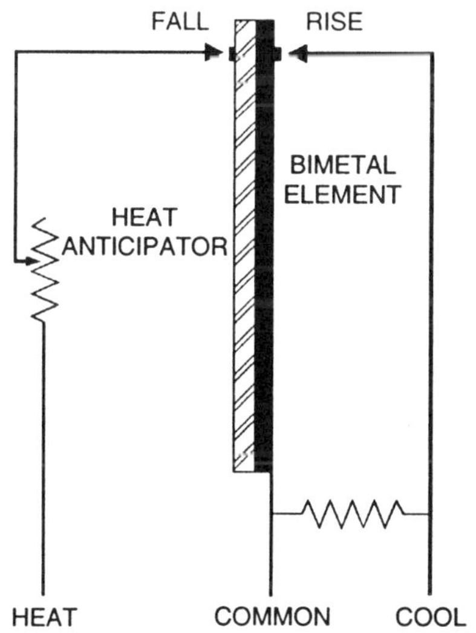

Figure 2-23. Schematic of a bimetal switch with heating and cooling anticipation.

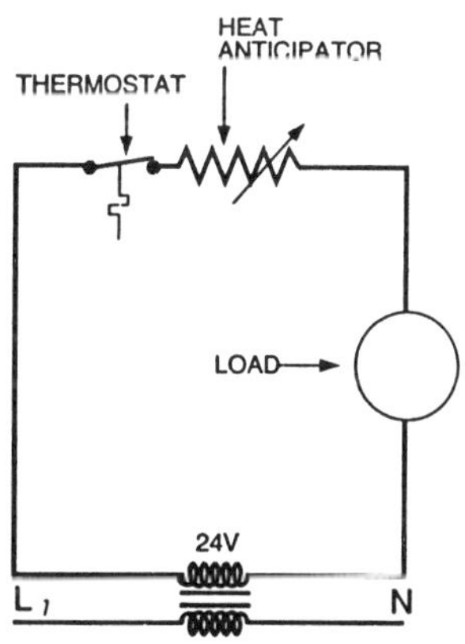

Figure 2-24. Placement of a heat anticipator.

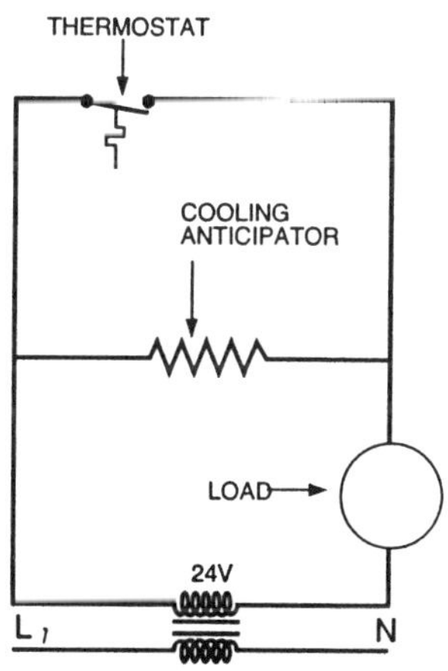

Figure 2-25. Wiring of the cooling anticipator.

PRESSURE CONTROLS

High- and low-pressure controls, Figures 2-26 and 2-27, are used as safety limits in cooling and refrigeration systems. Low-pressure controls are also used as motor cycling controls in refrigeration equipment.

As a limit protector, the high-side pressure control opens the compressor power circuit when above-normal pressure occurs. This prevents excessive pressure in the condenser. The low-side pressure control opens the power circuit to the compressor when normal pressure decreases. This prevents the air conditioner evaporator temperature from falling below the point at which frost forms on the coil. The low pressure control also cuts out the compressor upon loss of the refrigerant.

Pressure controls come with automatic or manual reset. Automatic reset restarts the system when the pressure returns to the cut-in setting on the control. The pressure switch used depends on the construction of the equipment and the preference of the manufacturer.

All safety controls should be set to the manufacturer's specification.

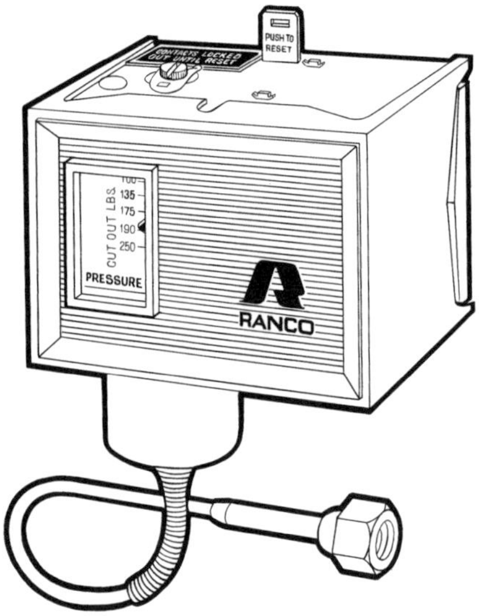

Figure 2-26. High pressure control. Courtesy, *Ranco Controls.*

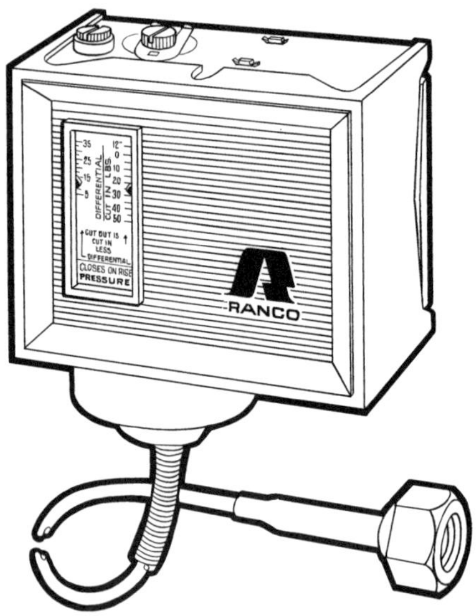

Figure 2-27. Low pressure control. Courtesy, *Ranco Controls.*

RESISTORS IN SERIES

Figure 2-28 shows four resistors in a series circuit: a 388 ohm center vertical drier, a pair of defrost heaters totaling 25 ohms and a 47,000 ohm defrost thermostat resistor. This circuit is typical of one found in a frost-free refrigerator.

The theory behind these resistors is complex; therefore, only some simple rules of thumb that explain how the resistors operate in this circuit will be discussed below.

High resistance produces little heat and low resistance produces maximum heat. Therefore, it is necessary to first take the 47,000 ohm defrost thermostat resistor and the 388 ohm center vertical drier out of the circuit before the two, 12.5 ohm, defrost heaters can achieve maximum heat production.

The heat produced by each series of resistors is directly proportional to the resistance of each heater. Therefore, the 47,000 ohm heater produces 99% of the heat in the four-resistor circuit. This amount of heat is insignificant, as it is only 0.3 watts. Basically, the 47,000 ohm resistor is only there to permit continuity testing through the defrost thermostat.

Since the current in a simple electric circuit is dependent on the applied voltage and the total resistance of the circuit, the total resistance of this circuit is

or $RT = R1 + R2 + R3 + R4$

$RT = 388 + 12.5 + 12.5 + 47{,}000 = 47{,}413$

Therefore, in solving this simple circuit, use the formula

$$\frac{E}{R} = I$$

$$\frac{115\text{v}}{47{,}413\text{ohms}} = 0\ .0024 \text{ amps}$$

To find the watts produced, use the formula

volts x amps = watts

115 volts x .0024 amps = 0 .276 watts

When the defrost thermostat closes, bypassing the 47,000 ohm resistor, the contacts on the defrost timer change position and bypass the 388 ohm center vertical drier. Now find the wattage for the defrost heaters.

$$\frac{E}{R} = I$$

$$\frac{115\text{v}}{25 \text{ ohms}} = 4.6 \text{ amps}$$

volts x amps = watts

115v x 4.6 amps = 529 watts

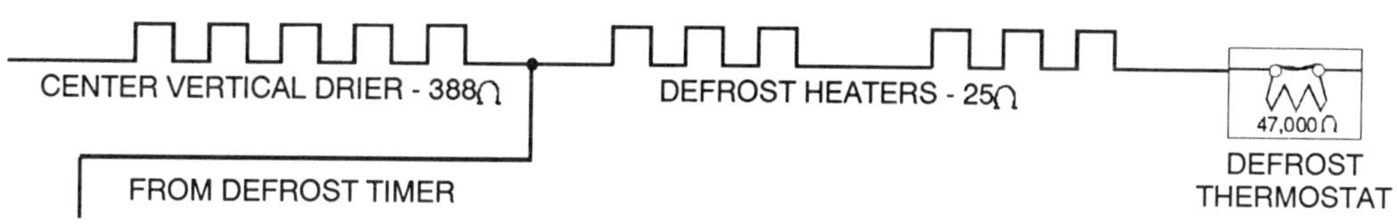

Figure 2-28. Resistors in series.

CAPACITORS

Capacitors are able to block direct current while passing alternating current. They can also fore electrical energy that can be used later. These actions improve the starting and running torque of split-phase motors and hermetic compressors used in air conditioning and refrigeration.

The unit of capacitance is the farad. Capacitors are usually rated in microfarads (μF). One microfarad equals one-millionth of a farad.

There are several types of capacitors in use. The most popular in the heating and cooling industry are the electrolytic (also called a start capacitor), Figure 2-29, and the oil-filled (also called a run capacitor), Figure 2-30.

The electrolytic capacitor is used to improve the starting torque of the motor. Due to the high current that must pass through the capacitor while starting, this capacitor is not recommended for continuous duty. This capacitor must also be removed from the circuit by a centrifugal switch, current relay, potential relay, or thermal relay once the motor attains its approximate rated speed.

The oil-filled capacitor is used to improve running torque. Its design and lower microfarad rating allow it to remain in the circuit, without causing overload or capacitor failure.

Figure 2-29. Electrolytic or start capacitor. Courtesy, *Sprague Electric Company.*

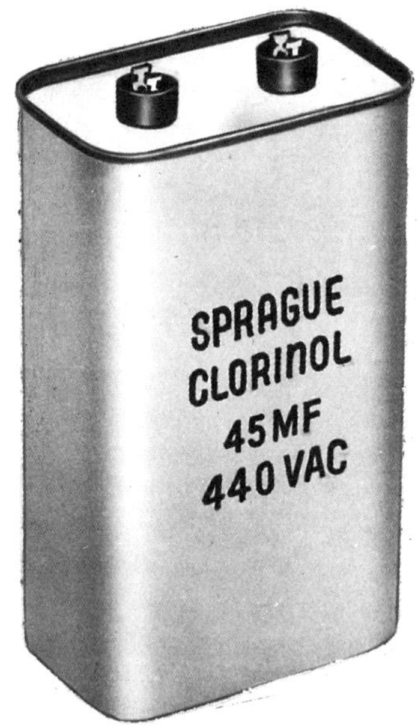

Figure 2-30. Oil-filled or run capacitor. Courtesy, *Sprague Electric Company.*

Section III

SOLID STATE ELECTRONIC CONTROLS

SOLID STATE FOR HVAC-R

It is the primary responsibility of the HVAC-R professional to maintain energy-consuming equipment at its highest operating efficiency. To accomplish this, highly accurate electronic devices, called solid state controls, are employed to provide precise control of energy use. The manufacturers' schematics in this chapter illustrate how these controls are used in heating, cooling and refrigeration equipment.

Solid state controls take the form of a complete circuit board, which is not repairable in the field. The Carrier Solid State Electronic Furnace Control Board, Figure 3-1, is an excellent example of a furnace control system. This furnace board controls all functions of the furnace.

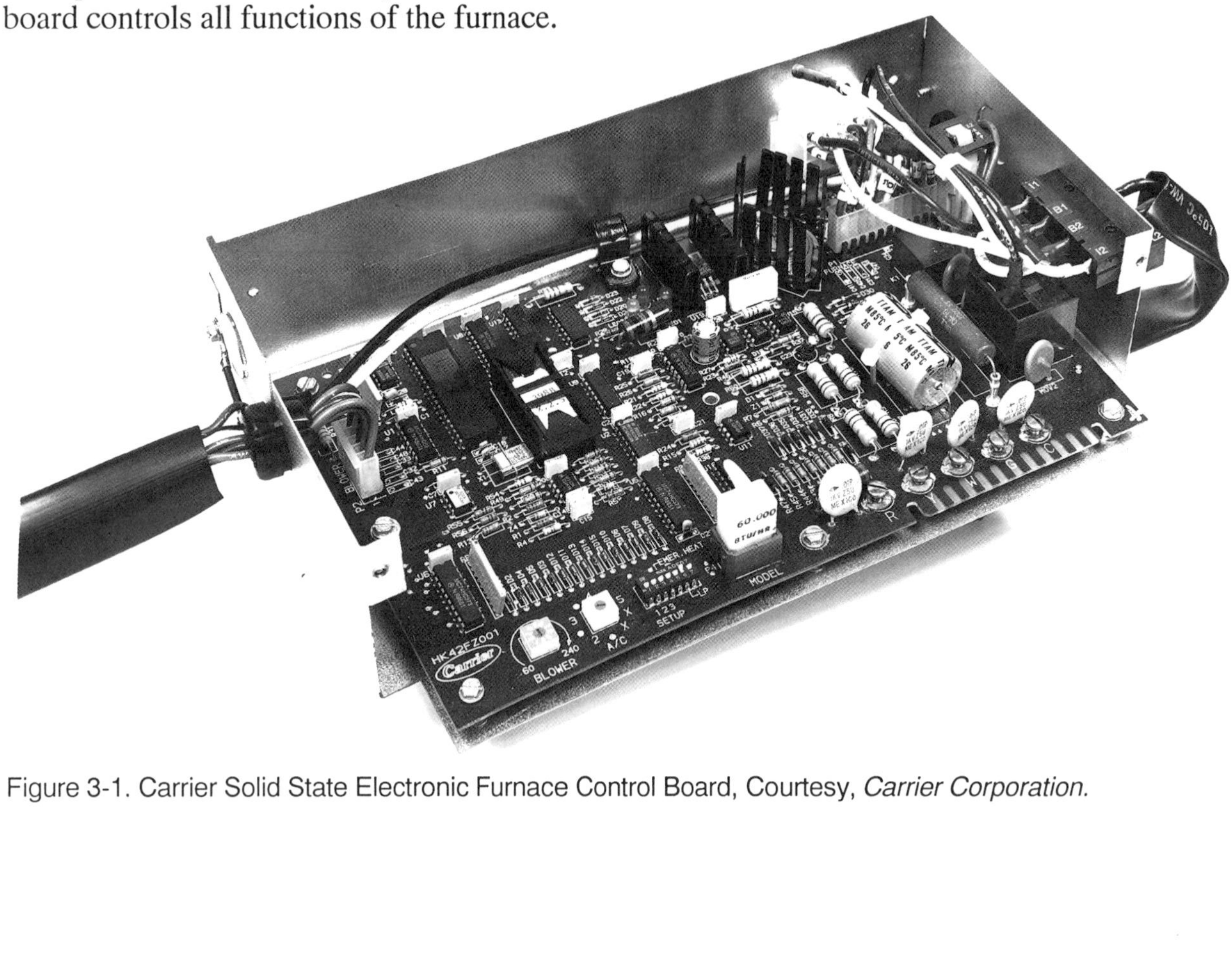

Figure 3-1. Carrier Solid State Electronic Furnace Control Board, Courtesy, *Carrier Corporation.*

ELECTRONICS SCHEMATICS

Figures 3-2 and 3-3 are examples of electronic schematics. The HVAC-R professional is not trained to read and understand this type of drawing, so instead, block diagrams are used. These block diagrams are illustrated on the following pages.

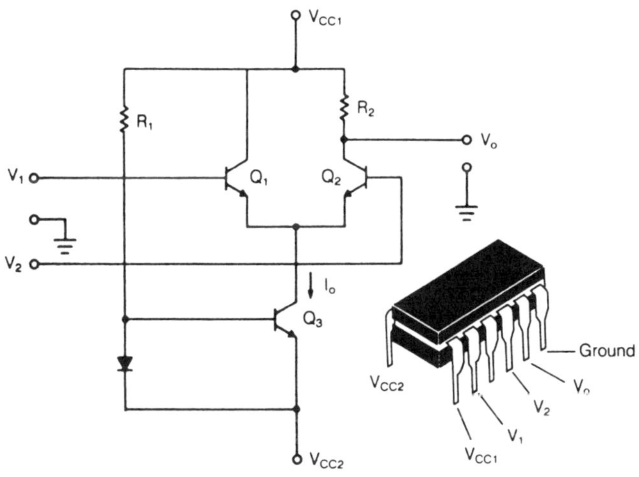

a. Schematic Diagram b. Integrated Circuit Package

Figure 3-2. Basic amplifier circuit. Courtesy, *Howard W. Samms, Incorporated.*

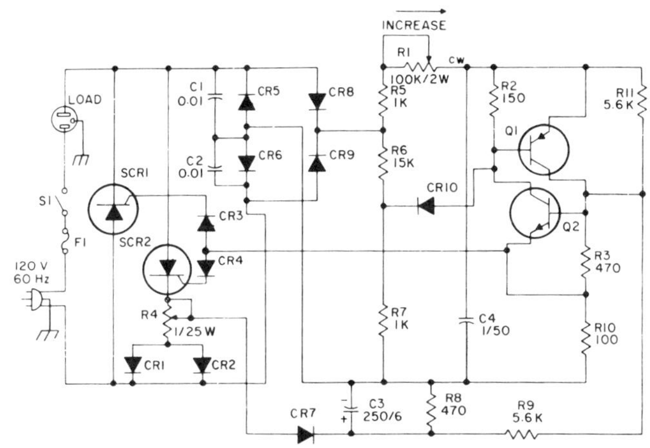

Figure 3-3. Schematic of a motor speed control. Courtesy, *RCA Corporation.*

Amplifier circuits, such as Figure 3-2, can be constructed using individual components. However, present manufactures make use of integrated circuit technology to make the circuit on one piece of silicon. This means the circuit can be purchased in a small plastic or ceramic package as shown.

FUNCTIONAL ANALYSIS

The complete understanding of a solid state electronic schematic requires training in the study of electronics and electronic circuitry. To assist the HVAC-R technician, a functional analysis is used to troubleshoot electrical problems. A functional analysis uses block diagrams to illustrate each progressive step (function) performed in an operation. Arrows are drawn between each block to show the next step and to illustrate the relationship of each component in a circuit. This method of analysis eliminates the need for mathematical calculations and an in-depth knowledge of electronics and electronic schematics.

A residential gas-fired heating system can be used to illustrate a functional analysis. A home heater is principally designed to generate heat for human comfort. Typically, this type of system is designed using familiar electromechanical and bimetal controls. The essential components of a gas-fired system are:

- Heat source
- Power source
- Fan-uniform heat distribution
- Thermostat-automatic control
- Safety control - overheating, fire prevention

Figure 3-4 illustrates how each of these components is connected to form a total functional analysis of the gas-fired heating system. The analysis shows the overall operation of the heating system in its entirety. However, it does not show how each component works on an individual basis. By studying this example, it iseasier to understand the operation of an electronic heating control system without having to rely on complicated electronic schematics.

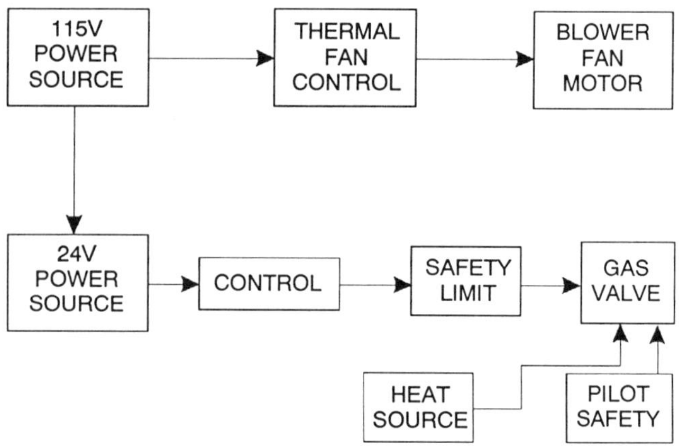

Figure 3-4. Functional analysis of a typical gas-fired heating system.

ELECTRONIC TIMERS

The electronic timer differs from the mechanical or thermal timer in four important ways:

- It only recognizes increments of one second (not minutes or hours)
- It must convert all time intervals into seconds (one hour equivalent to 3600 seconds)
- It operates on a frequency of 60 Hertz (timer actually counts the seconds)
- No clock motor or heating element is needed to drive the contact points open or closed

The electronic timer is a solid state, integrated circuit with no moving parts, Figure 3-5. Mounted in the integrated circuit (IC) are a number of distinct circuits. A block diagram, Figure 3-6 illustrates the functional operation of a simplified version of a timer.

Figure 3-5. Solid state timer. Courtesy, *Sealed Unit Parts Company, Incorporated.*

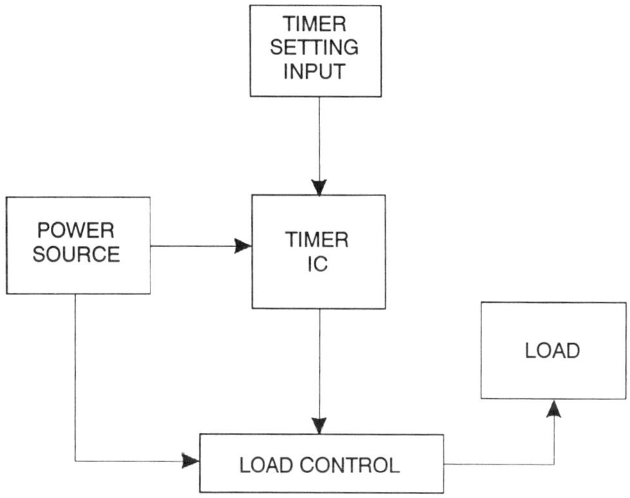

Figure 3-6. Block diagram of an electronic timer control.

Figure 3-7 illustrates the total electrical relationship of the IC timer to the other components in the heating system. From this drawing it becomes easier to visualize which components the timer affects and how they are affected.

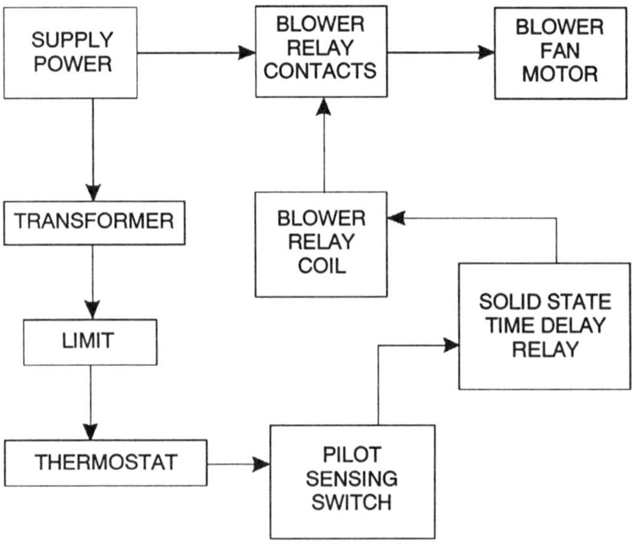

Figure 3-7. Block diagram of a solid state time delay relay.

Figure 3-8 is a manufacturer's schematic of a gas-fired system. Figure 3-9 is a photo of the actual blower control circuit board used in the system.

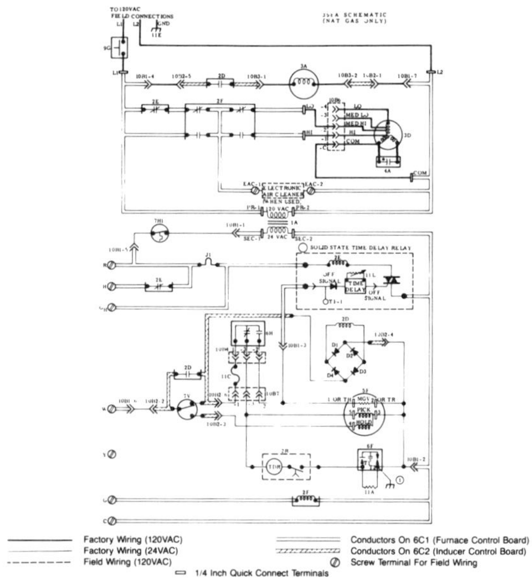

Figure 3-8. Manufacturer's schematic of a gas-fired system. Courtesy, *Bryant Air Conditioning/Heating*.

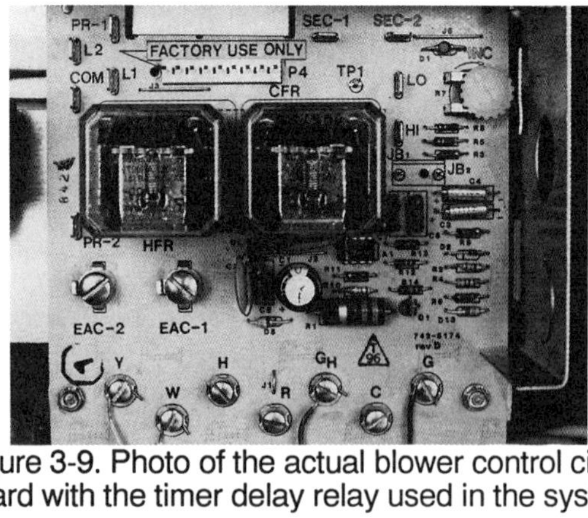

Figure 3-9. Photo of the actual blower control circuit board with the timer delay relay used in the system

ELECTRONIC THERMOSTAT

In addition to other operations, the electronic (solid state) thermostat performs all of the functions of the mechanical or bimetal thermostat. These other operations include: controlling a heating or cooling system so that it operates efficiently; saving energy; and delivering more comfort to the home.

Depending on the design and how it is programmed, the electronic thermostat can:

- Raise or lower boiler water temperature as the outdoor air temperature rises and falls.
- Increase or decrease room temperatures at different times of the day to save energy.
- Start the coffee maker and toaster, or check all windows and door locks to prevent burglaries.

Figure 3-10 is a block diagram that illustrates the function of each section of an electronic thermostat. Notice that all external information is fed to the microcomputer, which is the memory section.

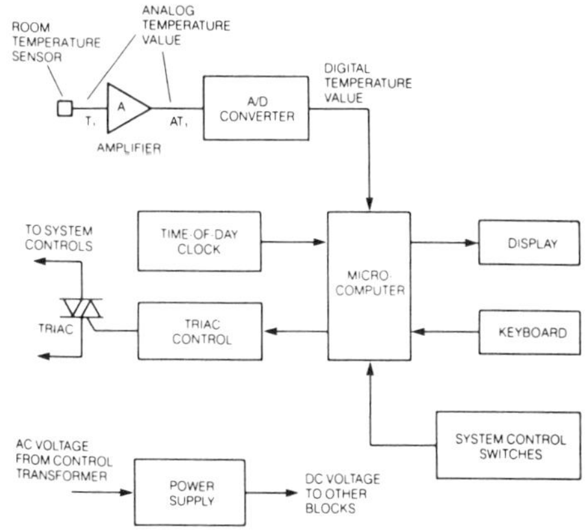

Figure 3-10. Block diagram of a typical electronic thermostat. Courtesy, *Howard W. Samms, Incorporated.*

Figure 3-11 illustrates a typical electronic thermostat control panel. All of the external input information is supplied from the control switches marked SYSTEM, FAN, and MODE. Additional information is supplied from the keyboard keys marked TIME, TEMP and NUMBERS, and from the solid state room temperature sensor. The system registers time and temperature readings on the LED display.

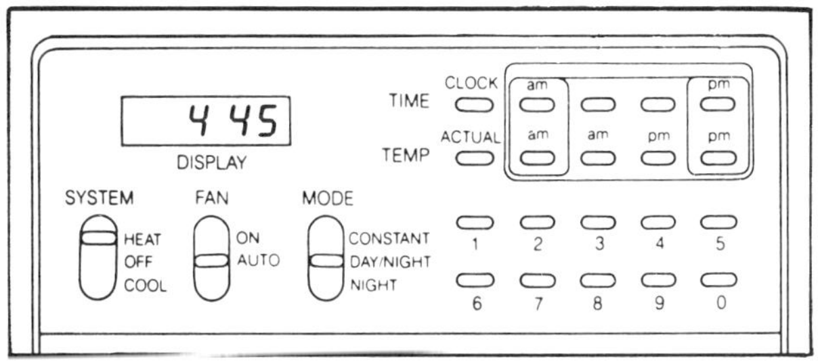

Figure 3-11. Electronic thermostat control panel.

FUNCTIONAL ANALYSIS OF A THERMOSTAT

Figure 3-12 demonstrates a functional analysis of a thermostat. The fundamental components of a thermostat are:

SIGNAL - The impulse received from a sensor.

REFERENCE - The information installed by the manufacturer to limit the high and low temperatures that control the operation of the system.

COMPARATOR- A circuit that compares two or more inputs and produces an output that is the result of the comparison.

CONTROL - The result of the comparator that is sent to the control with orders to open or close the circuit to the load.

From the illustration in Figure 3-12, it is easy to see how the four basic electronic elements affect each other and cause the thermostat to start or stop a cooling or heating system.

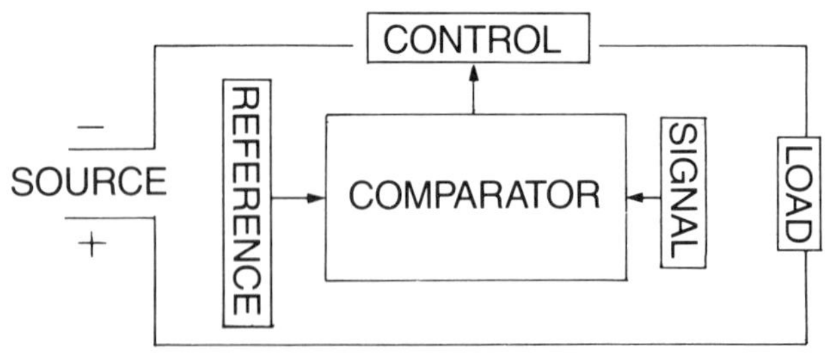

Figure 3-12. Functional analysis of a thermostat depict's basic components.

Figure 3-13 is the electronic schematic of a solid state time delay relay. This relay is used in a gas-fired hot air furnace to start the blower fan motor after a delay of approximately 60 seconds.

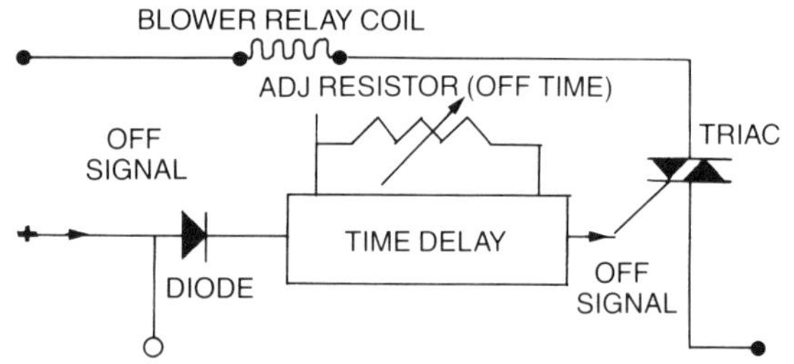

Figure 3-13. Schematic representation of a solid state time delay relay.

SOLID STATE RELAYS

Solid state relays can replace electromagnet current or amperage start relays. Figures 3-14 and 3-15 show two types of solid state relays. The relay in Figure 3-14 can be used on all compressors from 1/12 hp through 1/2 hp. This type can be used with or without a start capacitor. Installation is simple with only two wires to connect

The relay in Figure 3-15 can be used on all hermetic compressors from 1/12 hp through 1/3 hp, with or without a start capacitor. This model incorporates overload protection, which eliminates start winding overheating from locked rotor conditions. Installation requires the connecting of five wires, which are all marked for simplicity.

Figure 3-14. Solid state start relay. Courtesy, *Sealed Unit Parts Company Incorporated.*

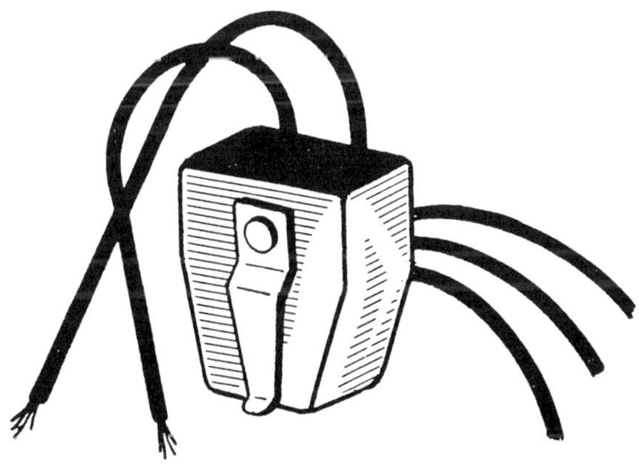

Figure 3-15. Solid state start relay with overload protection. Courtesy, *Sealed Unit Parts Company, Incorporated.*

START RELAY AND CAPACITOR

The solid state start relay and capacitor combination, Figure 3-16, can easily replace the electromagnet potential relay and start capacitor in all residential and commercial PSC air conditioners from 4,000 Btu through 60,000 Btu, 1/3 ton to 4 tons. AC voltage can be between 120 V and 288 V.

Similar to the potential relay, the internal solid state circuit is closed. When opened, it takes the start capacitor out of the circuit as the motor increases in speed. The solid state circuit can sense the proper voltage and current at the motor and act to block the capacitor at the right moment as the motor attains full speed. Installation is achieved in seconds with the connection made at the run capacitor terminals. This larger unit can increase starting torque up to 500%, making it excellent for hard starting conditions.

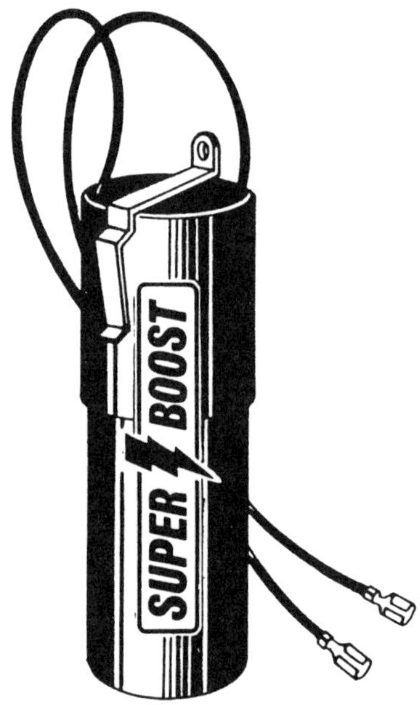

Figure 3-16. A solid state start relay capacitor. Courtesy, *Sealed Unit Parts Company, Incorporated.*

LOGIC MODULE

The logic module, Figure 3-17, responds to the demand signal of the thermostat. The module then examines the input from its four sensors (liquid, discharge, outdoor and defrost). From these inputs, the module determines if the heat pump and/or the auxiliary heat should operate. To save energy, the module accurately synchronizes the energy-saving features with the operation of the heat pump. Figure 3-18 is a block diagram of the Yorkguard logic module.

Figure 3-17. The Yorkguard logic module. Courtesy, *York International.*

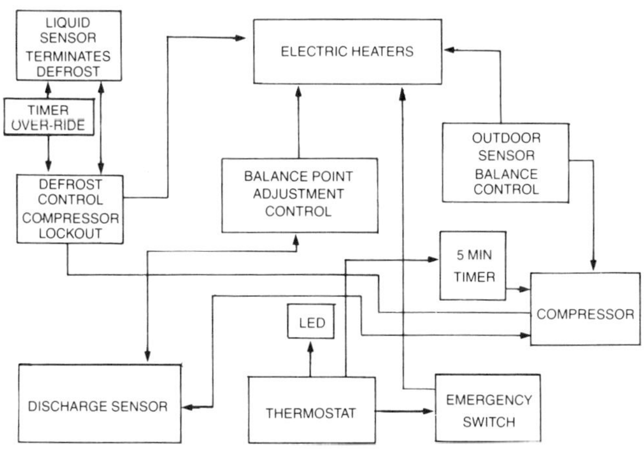

Figure 3-18. Block diagram of the Yorkguard logic module. Courtesy, *York International.*

A list of the logic module's components follows, along with a brief description of each.

- LIQUID SENSOR - terminates the defrost cycle when the liquid refrigerant rises to a predetermined temperature. Coupled with the defrost control, this sensor provides a timed override on the defrost cycle during gusty winds.
- DISCHARGE SENSOR - provides high discharge temperature protection for the compressor. Also protects against refrigerant loss or low air flow across the indoor or outdoor air coil.
- DEFROST CONTROL - provides defrost on demand. Also prevents nuisance defrosts caused by gusty winds. Locks out the heat pump if it attempts to defrost within five minutes after the last defrost has terminated.
- FIVE MINUTE TIMER - provides a five minute anti-cycle delay. Also prevents the compressor from restarting for five minutes after it has shutdown.
- OUTDOOR SENSOR - monitors the temperature of the outdoor air and permits the adjustable balance point thermostat to control the sequencing of the resistance heaters. It also shuts off the compressor when the outdoor air temperature drops below - 10 °F.
- ADJUSTABLE BALANCE POINT CONTROL - prevents the operation of the electric heaters above the balance point temperature. Also prevents supplemental heat operation when the heat pump can satisfy the building load.

ICE MAKER CONTROLS

Although freezing water is the only way to make ice, today's ice machine manufacturers have streamlined and automated the process. The automation process involves many steps which include:

- Removing air from the water to produce clear (uncloudy) ice
- Accurately measuring the proper quantity of water necessary for both ice production and harvesting
- Closely monitoring water temperatures to ensure correct harvest and defrost cycling

In addition, control of the overall machine operation is required when the ice storage bin fills or needs replenishing. Finally, the unit must have a safety shut-off.

Ice maker technology has advanced significantly since the early days. Besides having larger and faster ice making capabilities, modern ice makers are more reliable than their predecessors. This is due to the use of electronic control systems, Figure 3-19.

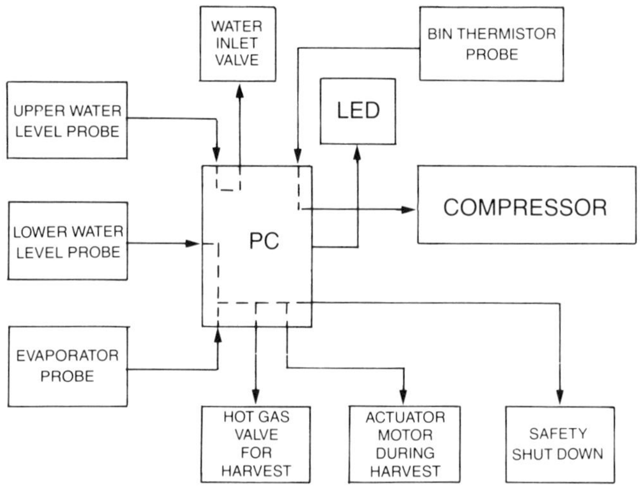

Figure 3-19. Block diagram showing the components of a typical electronic cuber.

The Kold Draft Division of Uniflow Manufacturing offers a troubleshooting device that illustrates what happens within an ice maker during each stage of its operating sequence, Figure 3-20. Used in conjunction with the step-by-step sequence of operation charts, this device guides service personnel though the troubleshooting process to assist in accurately diagnosing a malfunction. The device consists of a series of solid state probes connected to a printed circuit board control module. The unit monitors operation of an ice maker, discerns abnormal conditions and instantaneously acts to control them.

This unit identifies high/low water levels, as well as changes in resistance values of the evaporator probe due to fluctuating temperatures. Also, when the unit detects a temperature change in the evaporator, it performs three functions:

- Monitoring the cold water flow
- Controlling the actuator operation
- Regulating the required heat input

The troubleshooting unit also has a thermistor probe, which controls the ice bin level. This temperature probe senses the proximity of the ice at a given level in the storage bin and either actuates or terminates the ice making process. For example, if the ice level falls below the probe, the ice making process is activated. If the reverse occurs, the process is stopped.

Figure 3-20. Troubleshooting circuit board. Courtesy, *Kold Draft Division of Uniflow Manufacturing.*

Figure 3-20 shows the troubleshooting circuit board. The unit is color-coded and has switches that open or close a circuit, showing the operator how the current flows to each load. A series of LEDs indicates whether the ice maker is operating properly.

ELECTRONIC

The model 50 electronic relay, Figure 3-21, is designed to replace the use of sail switches, pressure switches and temperature switches. This relay interfaces a power humidifier with all types of forced warm air heaters. The installation only requires one simple connection to the blower motor circuit, as illustrated in Figures 3-22 and 3-23.

Figure 3-21. Current Sensing Relay - Model 50 (actual size of the relay). Courtesy, *Research Products Corporation. (Aprilaire)*

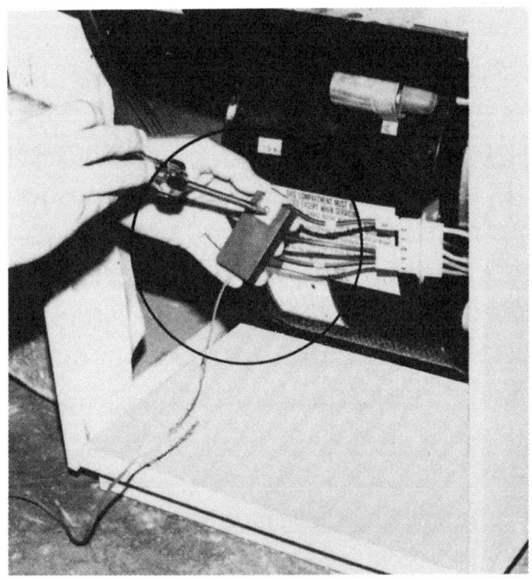

Figure 3-22. Illustrates the installation. The relay senses current flow to the blower motor, which starts the humidifier. The power to the heater motor does not have to be disconnected for installation.

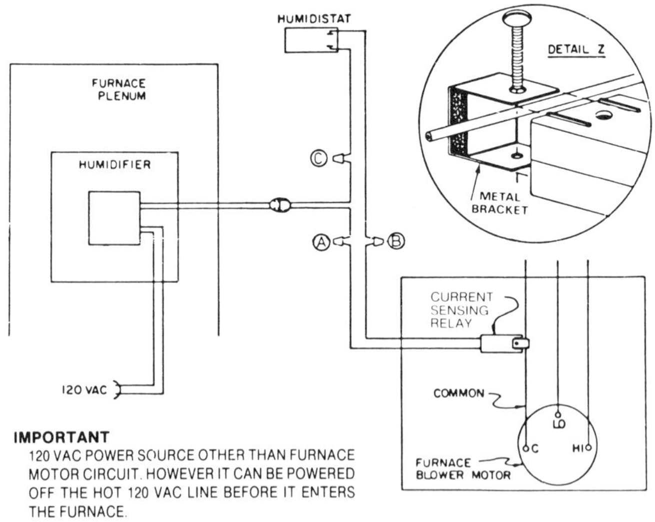

Figure 3-23. Wiring diagram for installation.

CONTROL FOR HOT SURFACE IGNITION

To eliminate the energy-using, standing pilot flame, a device called a hot surface ignition system is employed. This device first lights the pilot when the thermostat calls for heat. It then acts as a pilot safety after the pilot flame is proven. This control can be used on non-prepurge or prepurge gas-fired systems. On non-prepurge, the igniter energizes at once and remains energized for a period of either 17 or 34 seconds. The prepurge model has a time delay equal to the heat-up time selected before the ignitor is energized. At the end of the heat-up time, the gas valve opens. The ignitor is now de-energized and a sensor is energized. The system continues to operate as long as a flame is sensed or the thermostat or another safety control opens. At this time, both valves in the redundant gas valve close. Figure 3-24 is one example of this type of control.

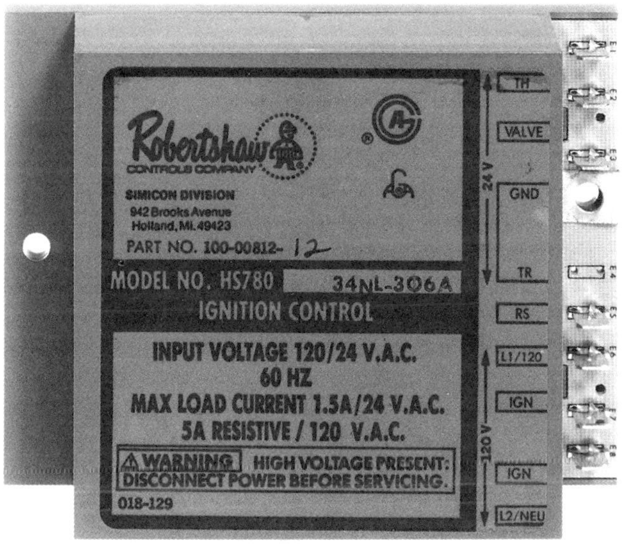

Figure 3-24. Hot surface ignition control. Courtesy, *Robershaw Controls Company.*

Section IV

TEST INSTRUMENTS

ELECTRICAL TEST EQUIPMENT

There are four factors in an electrical circuit:

Volts - a measurement of electrical pressure or force
Amperes - a measurement of the rate of electron flow
Ohms - a measurement of resistance to electron flow
Watts - a measurement of electrical power

A test instrument or meter measures each of these factors. These meters can measure one or more of the four functions and are available in all price ranges.

Meters are delicate instruments and must be handled with care. Extreme temperature, humidity changes, or rough handling, can destroy your most useful test equipment.

A safety warning accompanying one voltmeter/ohmmeter/milliammeter reads:

This instrument is designed to prevent accidental shock to the operator when properly used. However, no engineering design can render safe an instrument that is used carelessly.

A. Do not work alone when making measurements of circuits where a shock hazard can exist. Notify another person that you intend to make such a measurement.

B. Locate all voltage sources and accessible paths prior to making measurement connections. Check that the equipment is properly grounded and the right rating and type of fuse(s) is installed. Set the instrument to the proper range before power is applied.

C. Remember, voltages may appear unexpectedly in defective equipment. An open bleeder resistor may result in a capacitor retaining a dangerous charge. Turn off power and discharge all capacitors before connecting or disconnecting test leads to and from the circuit being measured.

D. For your own safety, inspect the test leads for cracks, breaks, or crazes in the insulation, probes, and connectors before each use. If any defects exist, replace the test leads immediately.

E. Do not make measurements in a circuit where corona is present. Corona can be identified by a pale-blue color emanating from sharp metal points in the circuit, the odor of ozone, and its sound.

F. Hands, shoes, floor, and workbench must be dry. Avoid making measurements under humid, damp, or other environmental conditions that could affect the dielectric withstanding voltage of the test leads or instrument.

G. For maximum safety, do not touch test leads or instrument while power is applied to the circuit being measured.

H. Use extreme caution when making measurements in a radio frequency (RF) circuit where a dangerous combination of voltages could be present, such as in a modulated RF amplifier.

I. Do not make measurements using test leads that differ from those originally furnished with the instrument.

J. Do not come into contact with any object that could provide a current path to the common side of the circuit under test or power line ground. Always stand on a dry insulating surface capable of withstanding the voltage being measured or that could be encountered.

K. The range or function switch should only be changed when the power to the circuit under measurement is turned off. This will provide maximum safety to the user, eliminate arcing at the switch contacts and prolong the life of the instrument.

Because the voltmeter is used to determine the presence of voltage, a potential danger to the user may exist. These precautions should be applied to the use of all test equipment that is used on a circuit with the main power supply on.

YOU NEED NOT FEAR ELECTRICITY
BUT YOU SHOULD RESPECT IT

THE VOLTMETER

Voltage is usually described as the force or pressure necessary to make the electrons flow. The voltmeter is the instrument used to measure the voltage, Figure 4-1.

When using the voltmeter to trace circuits, always place the voltmeter parallel to the circuit. Then, look for one of four readings:

1. No voltage across switch contacts indicates that the contacts are closed, Figure 4-2a.
2. A voltage reading across switch contacts indicates that the contacts are open, Figure 4-2b.
3. No voltage at the terminals of a motor or coil indicates that there is no voltage available to the motor or coil, Figure 4-2c.
4. A voltage reading at the terminals of a motor or coil indicates that voltage is available. If the motor or coil does not operate, then they are defective, Figure 4-2d.

Figure 4-1. Fluke 27 Multimeter. Courtesy, *John Fluke Manufacturing Company.*

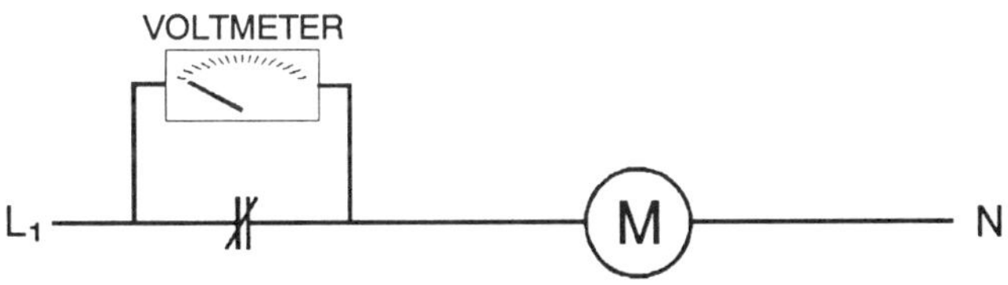

Figure 4-2a. No voltage indicates that the contacts are closed.

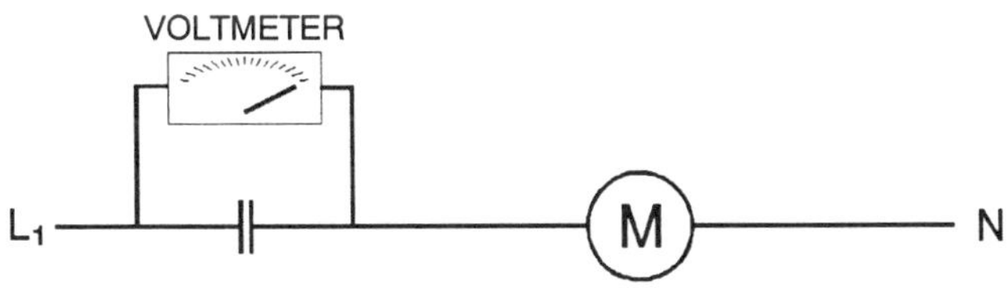

Figure 4-2b. Voltage indicates that the contacts are open.

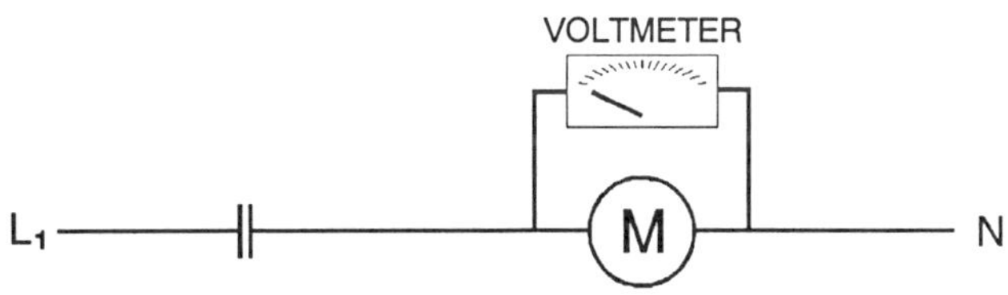

Figure 4-2c. No voltage indicates that there is no voltage available to the motor.

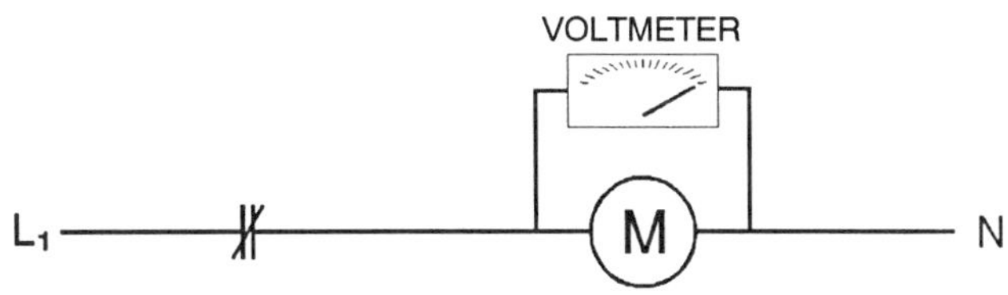

Figure 4-2d. Voltage indicates that there is voltage available to the motor.

AMMETERS

The ampere, which is the rate of flow of electrons or current, is measured with an amp meter, or ammeter, Figure 4-3. The ammeter must be connected in series with the load, which can be time consuming.

The clamp on the digital ammeter, Figure 4-5, quickly reads the current draw of an electrical load. In this respect, it is similar in use to the Amprobe clamp on an analog ammeter, Figure 4-4.

A current draw that is 10% above or below the rated amperage of a load is permissible. Greater variations than 10% indicate that a problem is developing or has already developed.

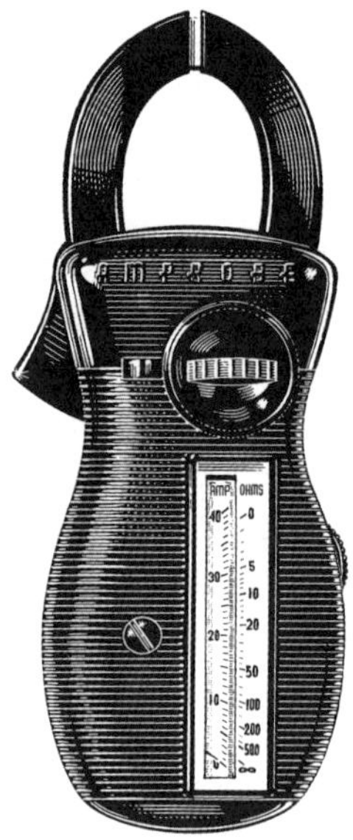

Figure 4-4. Amprobe Volt, Amp, Ohmmeter. Courtesy, *Amprobe Instrument.*

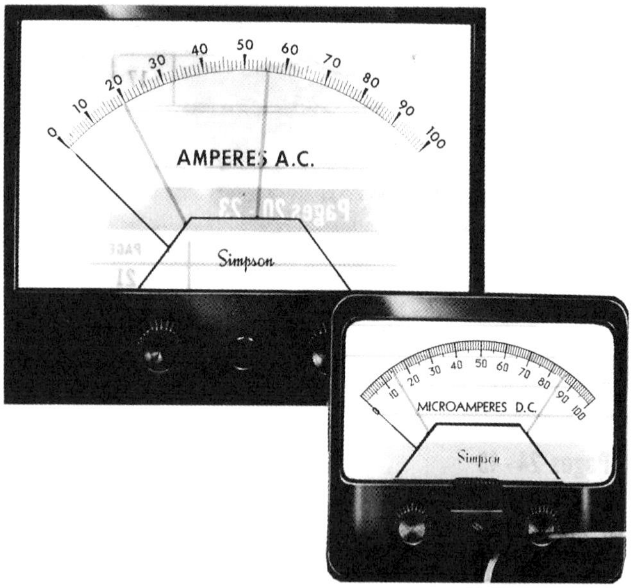

Figure 4-3. Ammeter. Courtesy, *Simpson Electric Company.*

Figure 4-5. TIF 1000 LED Digital Clamp-On Ammeter. Courtesy, *TIF Instruments, Incorporated.*

THE OHMMETER

WARNING: The ohmmeter can only be used when the power supply is disconnected. The ohmmeter operates from a battery within the instrument rather than from line voltage.

Resistance to electron flow, measured in ohms, is present is all conductors.

The ohmmeter can test capacitors, find common, run and start in a single-phase compressor motor, and act as a continuity tester when checking fuses, Figure 4-6.

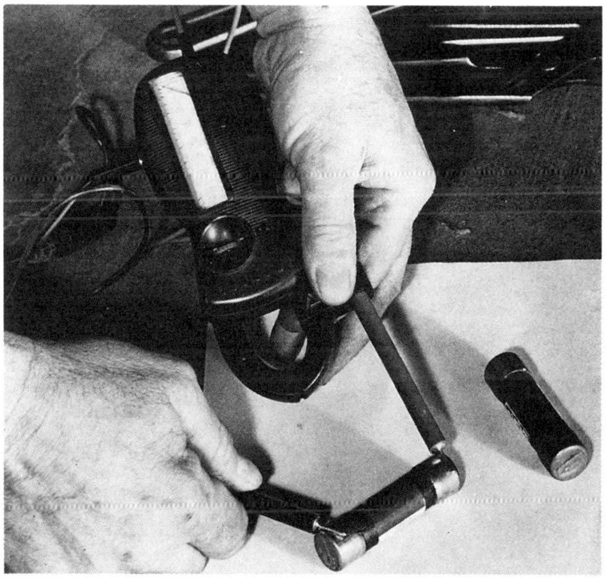

Figure 4-6. Using an Amprobe as an ohmmeter to check fuses.

TESTING AC CAPACITORS

WARNING: Always discharge capacitors before making any test and before removing the capacitor from the circuit. A fully charged capacitor could cause a burn if discharged through your body.

One method used for quick testing on the job is the ohmmeter test. First, set the meter to the highest range. Then:

1. If the capacitor is shorted, there will be a low resistance reading. The meter needle could move to zero.
2. If the capacitor is open, the resistance will read infinity. The needle will not move.
3. If the capacitor is good (having some capacitance), the needle will deflect rapidly while the capacitor charges itself from the battery in the ohmmeter. The needle will then drift slowly back to infinity as the charge bleeds away.

The capacitor must be discharged each time this test is performed.

The ohmmeter test is not conclusive, but it does indicate whether the capacitor is shorted or open or not shorted or open. The test does not detect a weak capacitor.

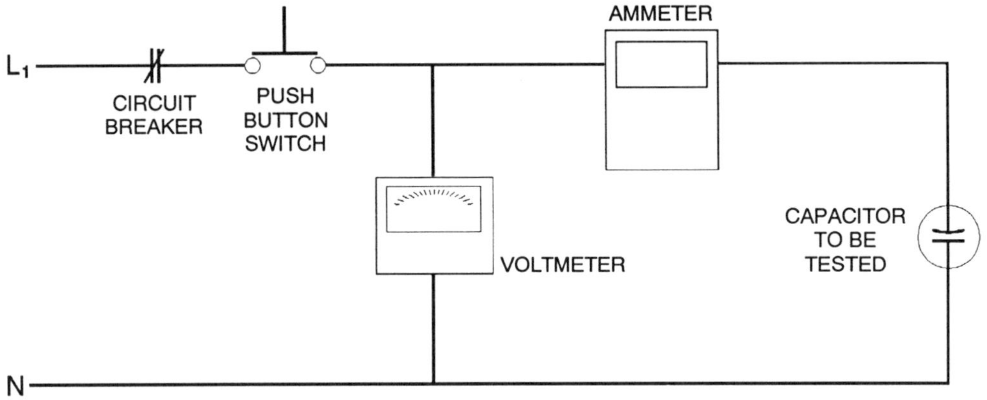

Figure 4-7. Testing a capacitor with a voltmeter and an ammeter that have been wired together.

Several companies have developed test equipment that accurately measures capacitance, or microfarads (μF). The tester consists of an ac voltmeter and an ac ammeter wired together as shown in Figure 4-7. While the capacitor is in the test circuit:

1. Test the capacitor for not more than 10 seconds.
2. Note both the ammeter and voltmeter readings.
3. Calculate the capacity using this formula

$$\frac{2650 \text{ x amps}}{\text{volts}} = \mu F$$

This method will give a good approximation of capacity, within $\pm$10%.

1. If capacity is much lower than rated μF, the capacitor is weak and overage.
2. If capacity is much higher than rated μF, the capacitor is probably developing an internal short.

Remember that capacitors are affected by temperature. All tests must be conducted at 25 °C (77 °F). A capacitor that is too hot or too cold will not give accurate test results.

IDENTIFYING MOTOR TERMINALS

If motor terminal markings are illegible, use the ohmmeter to measure the resistance between the three terminals that you have labeled 1,2 and 3, as shown in Figure 4-8. You might find that the resistance between 1 and 2 is 5 ohms, between 2 and 3 is 15 ohms and between 1 and 3 is 20 ohms. The highest reading indicates that you have measured the resistance of both the start and run windings in series. Therefore, 1 and 3 are obviously run and start and 2 is common. Since you know that the start winding has the higher resistance, 3 is start and 1 is run.

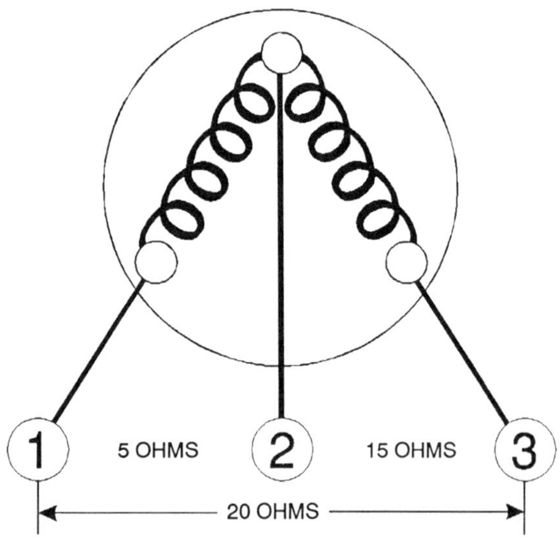

Figure 4-8. Identifying Motor Terminals.

POWER FACTOR METER

The power factor (PF) of a motor is the ratio (expressed as a decimal or percentage) of true power (measured in watts) to the apparent power (measured in volts x amps, VA). The power factor is never higher than 1, or 100%, and rarely lower than 0.7, or 70%. Example: A watt meter connected to a typical inductive alternating current (ac) load reads 1800 watts. The VA reading for the same load is 2000 watts. The ratio of the true power (1800 watts) to the apparent power (2000 watts) power factor is 90%. This indicates that 90% of the power is in phase and 10% of the power is out of phase. The power out of phase does no work, but it does create heat within the circuit at a cost. For this test, a power factor meter giving direct readings in PF is used, Figure 4-9. The watt meter, Figure 4-10, can also be used to determine motor power factor.

Figure 4-9. A direct reading power factor meter. Courtesy, *Amprobe Instrument Company.*

Figure 4-10. Model DU-6060 watt meter, Courtesy, *Extech Instruments, Incorporated.*

Section V

CREATING ELECTRICAL SCHEMATICS

THE SCHEMATIC LAYOUT SHEET

The schematic layout sheet is used for all 10 examples in this section. Each example requires a different schematic layout sheet. These perforated layout sheets are located in Appendix A.

The vertical line at the left, L_1 is a hot line. The vertical line at the right is the L_2, or N, neutral line. Together they are L_1 - L_2 240-volt, single-phase, alternating current (ac), or L_1 - N 120-volt, single-phase, alternating current (ac).

The examples themselves will start with very simple circuits and progress to rather complex electrical systems. They are designed to give you increasing confidence in your ability to understand and work with a wide range of common electrical control circuits.

At the end of each example is a series of questions directed to the finished example. The questions are designed to test your knowledge as an electrical troubleshooter. The answers to these questions can be found in Section VII.

Before the electrical schematic can be drawn, you must first do three things.

1. DEFINE THE FUNCTION OF THE EQUIPMENT

Ask yourself: what is this equipment and what are its circuits designed to do? The function of the basic forced-air furnace is to generate heat for your personal comfort. The essential components are a source of heat and a fan to distribute this heat. Do you want uniform comfort? Of course you do, so you control the fan with a plenum switch. Do you want to regulate the heat automatically? Of course you do, so you add a thermostat. Now the function of the furnace has to be redefined, based on your additional requirements. Your furnace is no longer just a means to generate heat, it can now uniformly and automatically generate and distribute heat. Still, if you go one step further, adding a little more equipment and two more electrical circuits, this same furnace can both heat and cool, to give you year-round personal comfort.

This is the way you must define the function of the equipment, because function determines the components to be used in your next step.

2. LIST THE ELECTRICAL COMPONENTS

Section II gave you a brief description of some of the controls you will be employing in these examples. How do you know which ones to use?

Once you perform Step 1, review the function of the equipment. In this case, the equipment is an automatically- and uniformly-controlled, gas-fired, forced-air heating and cooling system operating on a standard, domestic power supply with low voltage controls.

Based on this definition, here are the components and sources you need:

STANDARD DOMESTIC
POWER .... *120V, 60 HZ, 1 ϕ , AC*
GAS-FIRED .... *GAS VALVE*
FORCED AIR .... *FAN MOTOR*
UNIFORM COMFORT .... *PLENUM SWITCH*
AUTOMATIC HEATING
AND COOLING .... *HEATING AND COOLING THERMOSTAT*
LOW VOLTAGE CONTROL .. *TRANSFORMER*
COOLING .... *COMPRESSOR AND RELAY*

3. GO STEP-BY-STEP THROUGH THE ELECTRICAL SEQUENCE

In other words, ask yourself, "What happens next?"

Right now, if you were asked, "How do you get cold air?" You might reply, "You turn down the thermostat until the fan goes on. Then, you get cold air."

Later, when you have mastered the schematic, you will know that the process of producing cold air requires not one, but a number of steps in sequence. And, you will know that one step is separated from the other by a microsecond time interval, even though the entire sequence seems to occur instantly.

Let's apply this 3-step approach to our first example, an air-conditioning compressor.

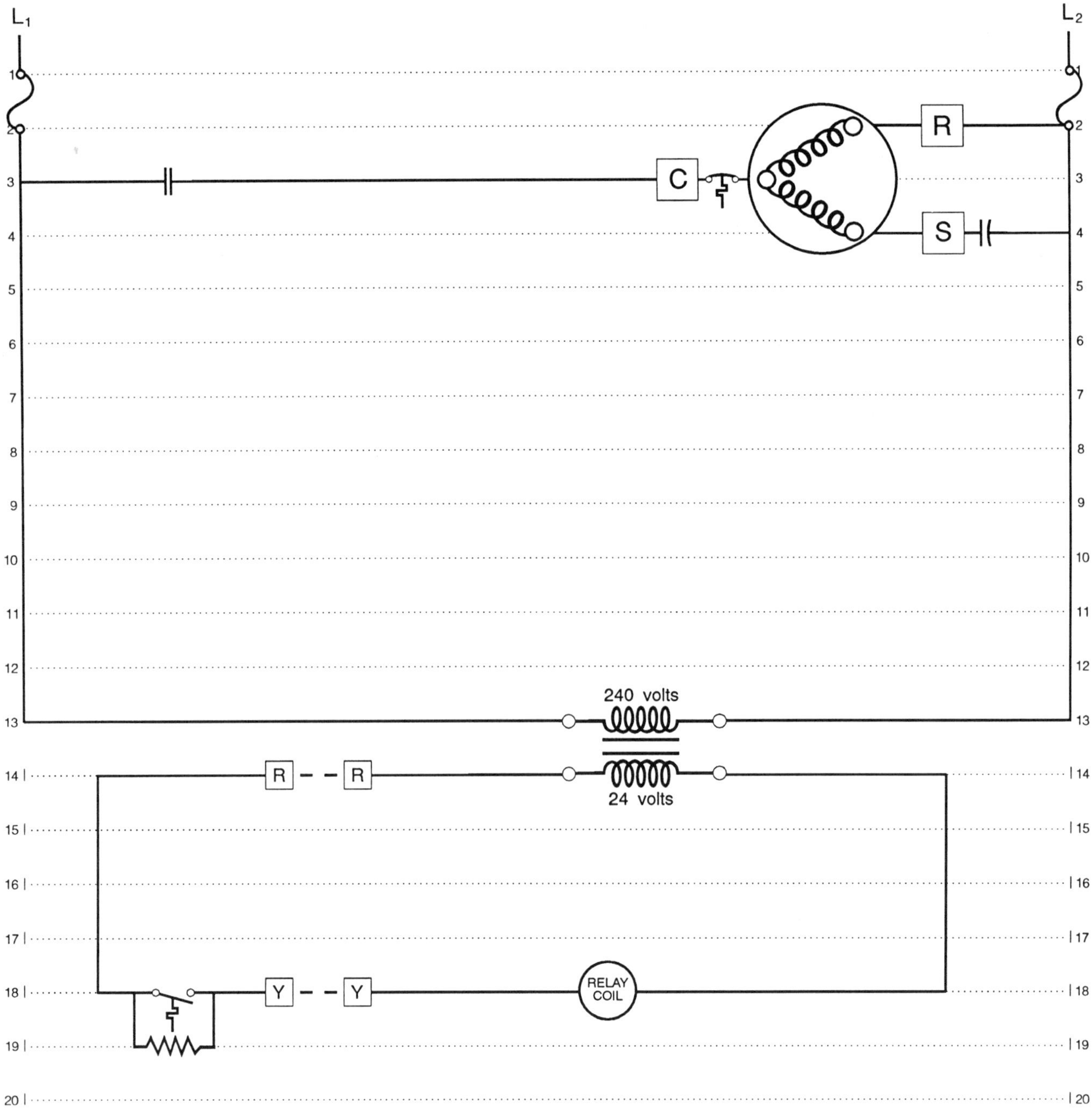
L1
L2
C
R
S
240 volts
24 volts
R
R
Y
Y
RELAY COIL

EXAMPLE 1
AIR CONDITIONING COMPRESSOR

1. DEFINE THE FUNCTION OF THE EQUIPMENT

This is an internally-protected compressor, automatically operated by a low-voltage thermostat, from a 24-V source.

2. LIST THE ELECTRICAL COMPONENTS

- The standard electrical source is 240 V, 60 cycle (60 Hz), single-phase (1 ϕ), alternating current (ac).
- Fuses are required by the Safety Code.
- The internal protector is a standard feature of all compressors. Actually, it is wired inside the shell of a hermetically-sealed compressor. Schematically, it is shown in series with the common leg of the two-circuit compressor motor.
- The compressor may have one of a number of starting circuits. In this example, the compressor employs a permanent-split capacitor (PSC) motor and has a RUN capacitor S in the motor starting circuit. The use of a PSC motor tells you that the compressor requires low starting torque.
- A contactor is used to start the compressor and is wired in series with the internal protector. The coil for the contactor is wired into the 24-V circuit.
- A step-down transformer (from 240 V to 24 V) is required for the low-voltage circuit.
- A cooling thermostat, operating on 24 V, is required for the control circuit.
- Field wiring from the thermostat to the compressor contactor coil is required.
- Terminal R on the thermostat is connected to terminal R on the transformer. Terminal Y on the thermostat is connected to Y on the contactor.

3. GO STEP-BY-STEP THROUGH THE ELECTRICAL SEQUENCE

- Line power must be supplied to the high (240 V) side of the transformer.
- When the thermostat closes on a temperature rise, the circuit is completed between terminals R and Y.
- This induces a 24-V current in the low side of the transformer and energizes the compressor relay coil.
- The normally-open contacts in the 240-V circuit, close to supply current, through the internal protector, to the compressor motor.

4. DRAW THE SCHEMATIC

Using the layout sheet:

STEP 1. Insert a fuse in L_1 and L_2 between lines 1 and 2.

STEP 2. Draw the transformer between lines 13 and 14 and connect the 240-V side of the transformer to L_1 and L_2 along line 13.

STEP 3. On line 14, draw terminal [R] and connect it, with low voltage field wiring - - -, to another [R] terminal on the 24-V side of the transformer. Connect the transformer to [R].

STEP 4. On line 18, draw terminal [Y] and connect it, with field wiring - - -, to another terminal [Y]. Draw a contactor coil (RELAY COIL) and connect it to [Y] and the 24-V side of the transformer.

STEP 5. On line 18, draw a normally-open thermostat and connect it to terminals [R] and [Y].

STEP 6. On line 19, draw a resistor -WW- and connect it to line 18 to create a parallel circuit with the thermostat. This is the cooling anticipator.

STEP 7. On line 3, connect the contactor contacts —||— to L_1. Connect the contacts to the common lead [C] of the compressor motor. Position the internal protector between [C] and the junction of the start and run windings to indicate that it is an integral part of the common lead to the motor.

STEP 8. Connect the RUN winding [R] to L_2 along line 2. Connect the motor start winding to the RUN capacitor [S]—||— on line 4 and the ground side of the capacitor to line L_2.

The tear-out schematic layout sheet can be found on page 225.

Step 1

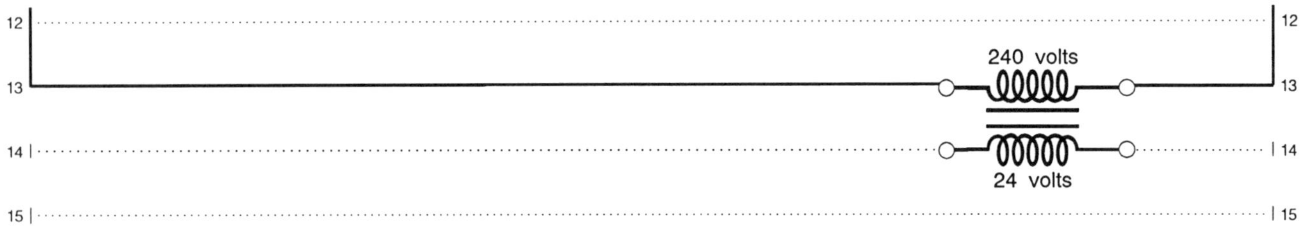

Step 2

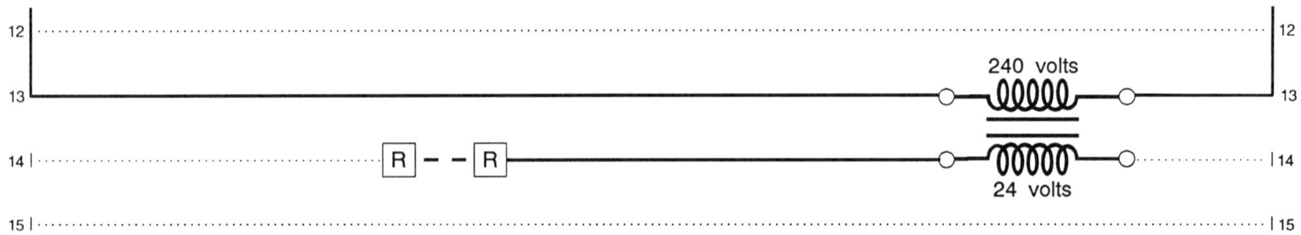

Step 3

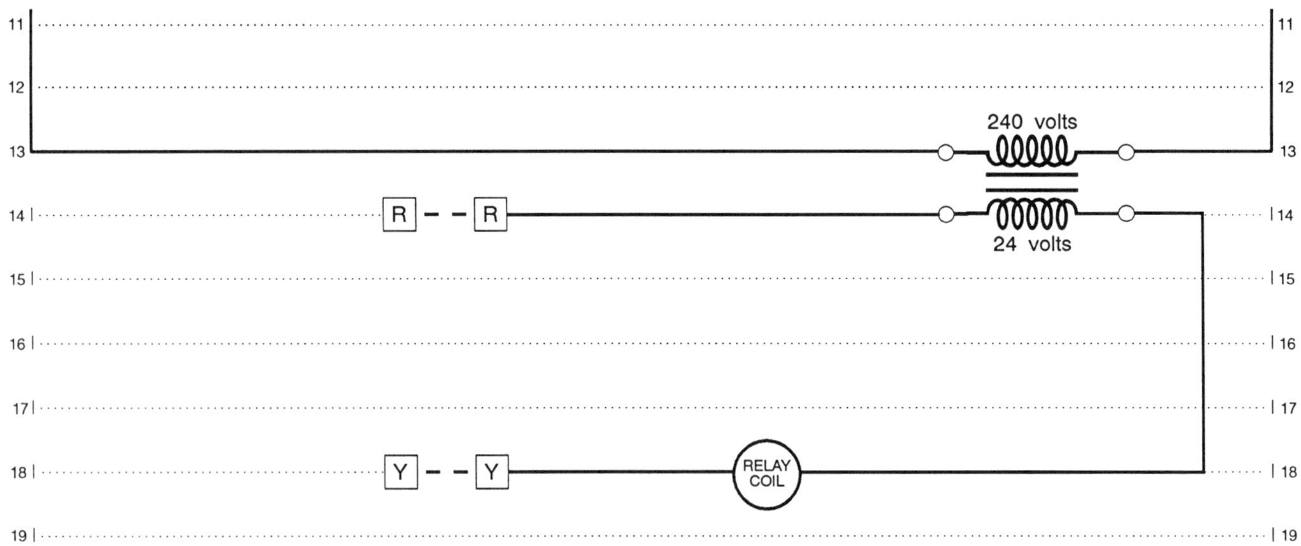

Step 4

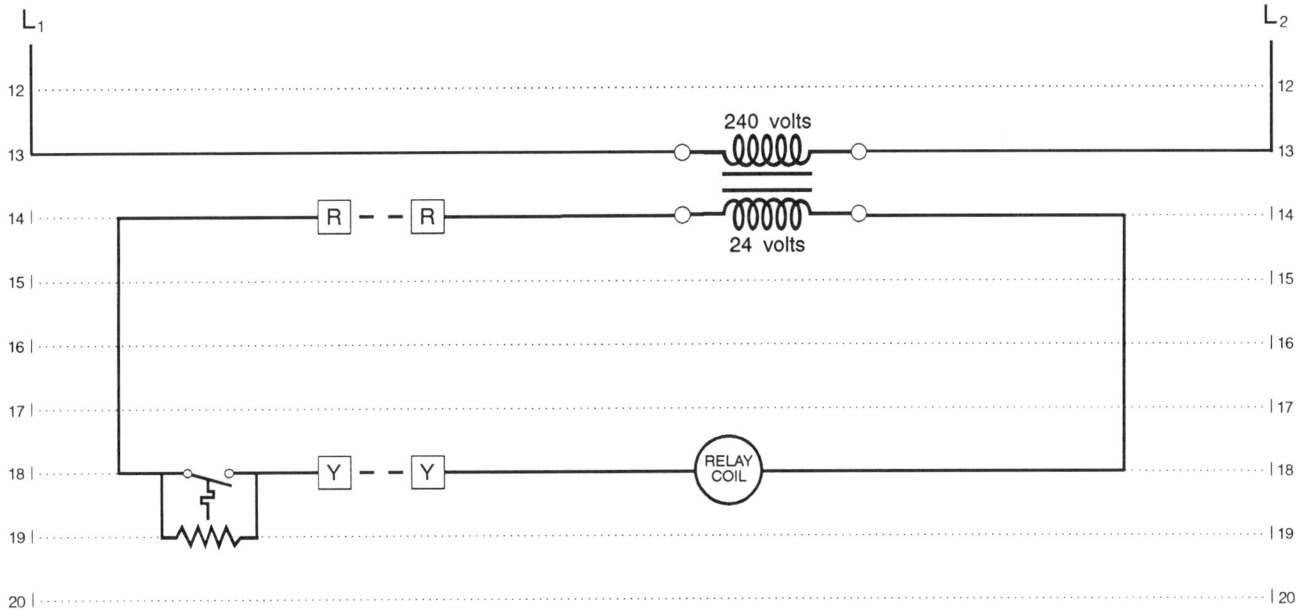

STEPS 5 AND 6

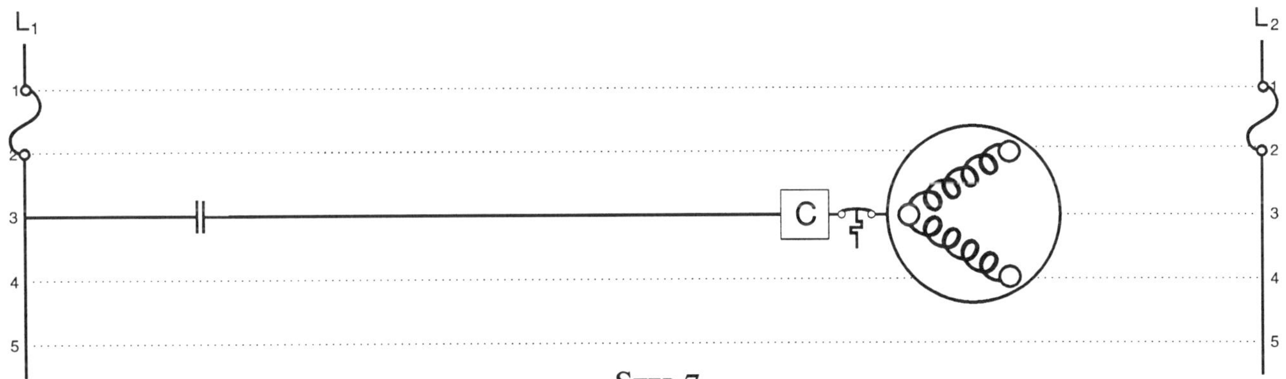

STEP 7

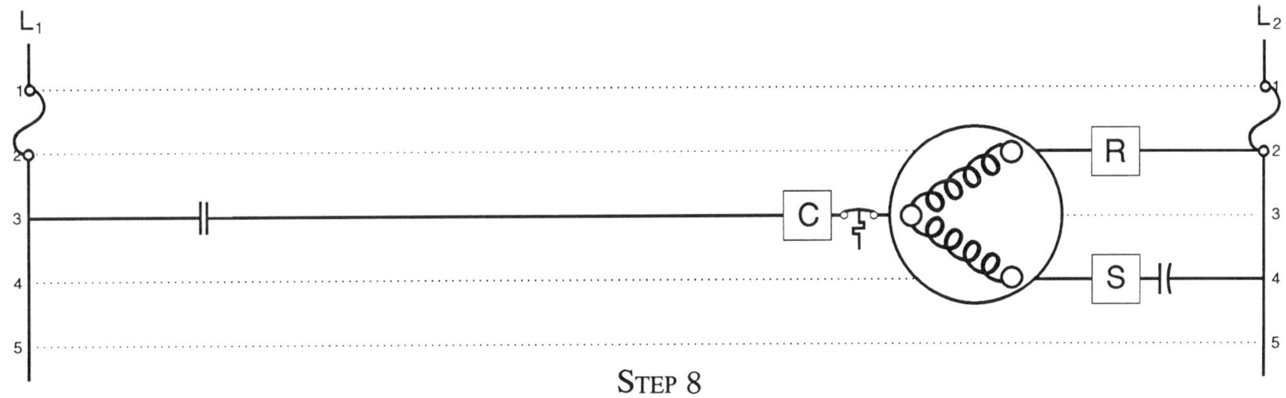

STEP 8

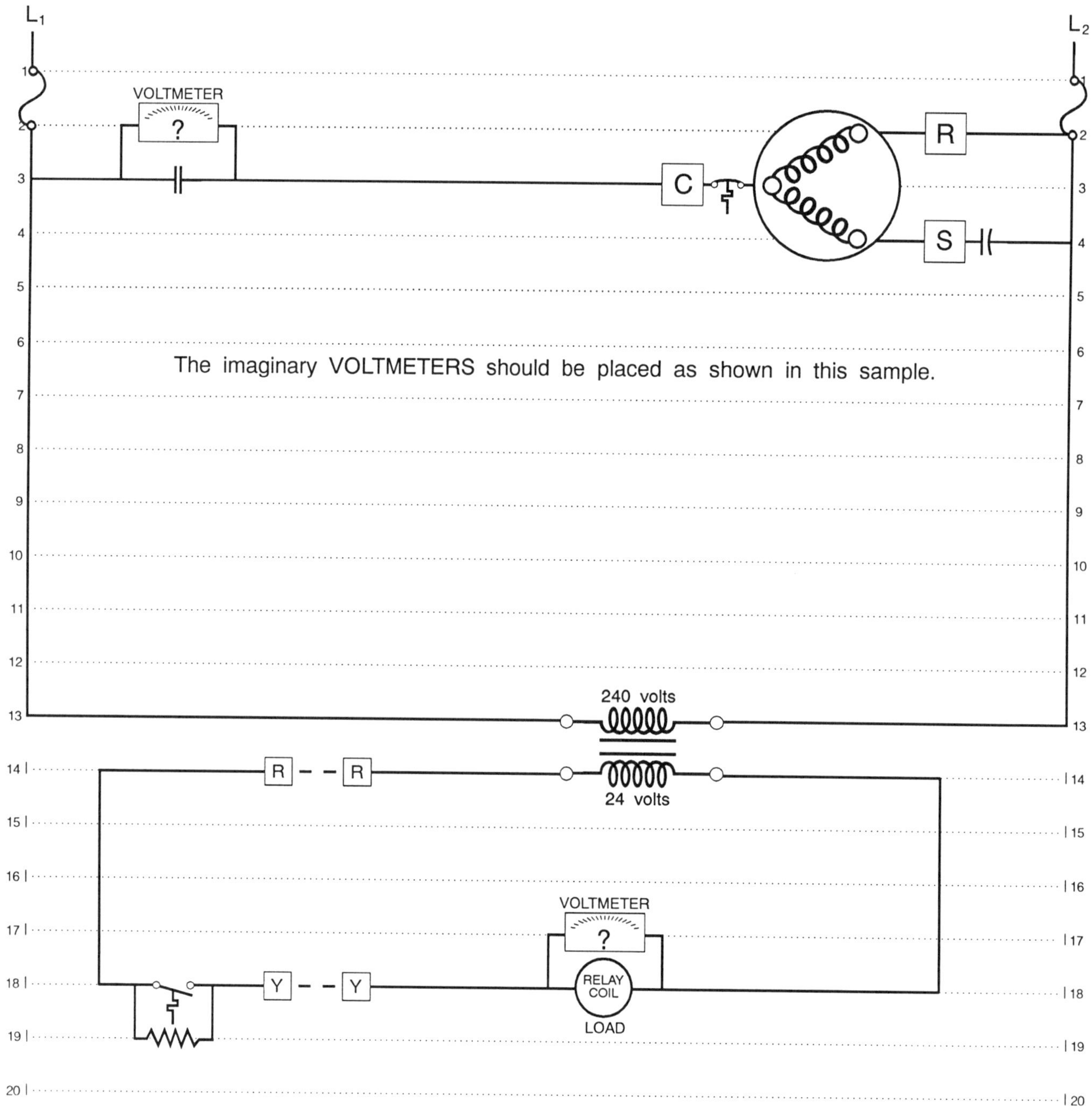

L1
L2
VOLTMETER
?
C
R
S
The imaginary VOLTMETERS should be placed as shown in this sample.
240 volts
24 volts
R
R
VOLTMETER
?
Y
Y
RELAY COIL
LOAD

QUESTIONS FOR EXAMPLE 1

Using an imaginary voltmeter and the sample drawing, place one probe on each side of the contact or load given, as pictured. The voltages that you find are to be entered on the answer sheet. Always read the conditions given for each question before each reading is to be taken.

(WHAT VOLTAGE WILL YOU READ?)

Q-1, Place the voltmeter probes on either side of the thermostat contacts [R] and [Y]. What is the voltage with the thermostat contacts [R] [Y] open?

A-1, _____V

Q-2, Thermostat terminals, line 18, with the compressor operating?

A-2, _____V

Q-3, Relay coil, line 18, with the thermostat contacts open?

A-3, _____V

Q-4, Relay coil, line 18, with the compressor operating?

A-4, _____V

Q-5, Relay coil, line 18, with the thermostat contacts closed?

A-5, ______V

Q-6, Transformer primary, line 13, with the thermostat contacts open?

A-6, ______V

Q-7, Transformer primary, line 13, with the thermostat contacts closed?

A-7, ______V

Q-8, Transformer secondary, line 14, with the thermostat contacts closed?

A-8, ______V

Q-9, Transformer secondary, line 14, with the compressor operating?

A-9, ______V

Q-10, The compressor terminals [R], (line 2) and [S], (line 4) with the compressor operating?

A-10, ______V

Q-11, The compressor terminals [R], (line 2) and [C], (line 3) with the thermostat in the open position?

A-11, ______V

Q-12, The compressor terminals [R], (line 2) and [C], (line 3) with the compressor operating?

A-12, ______V

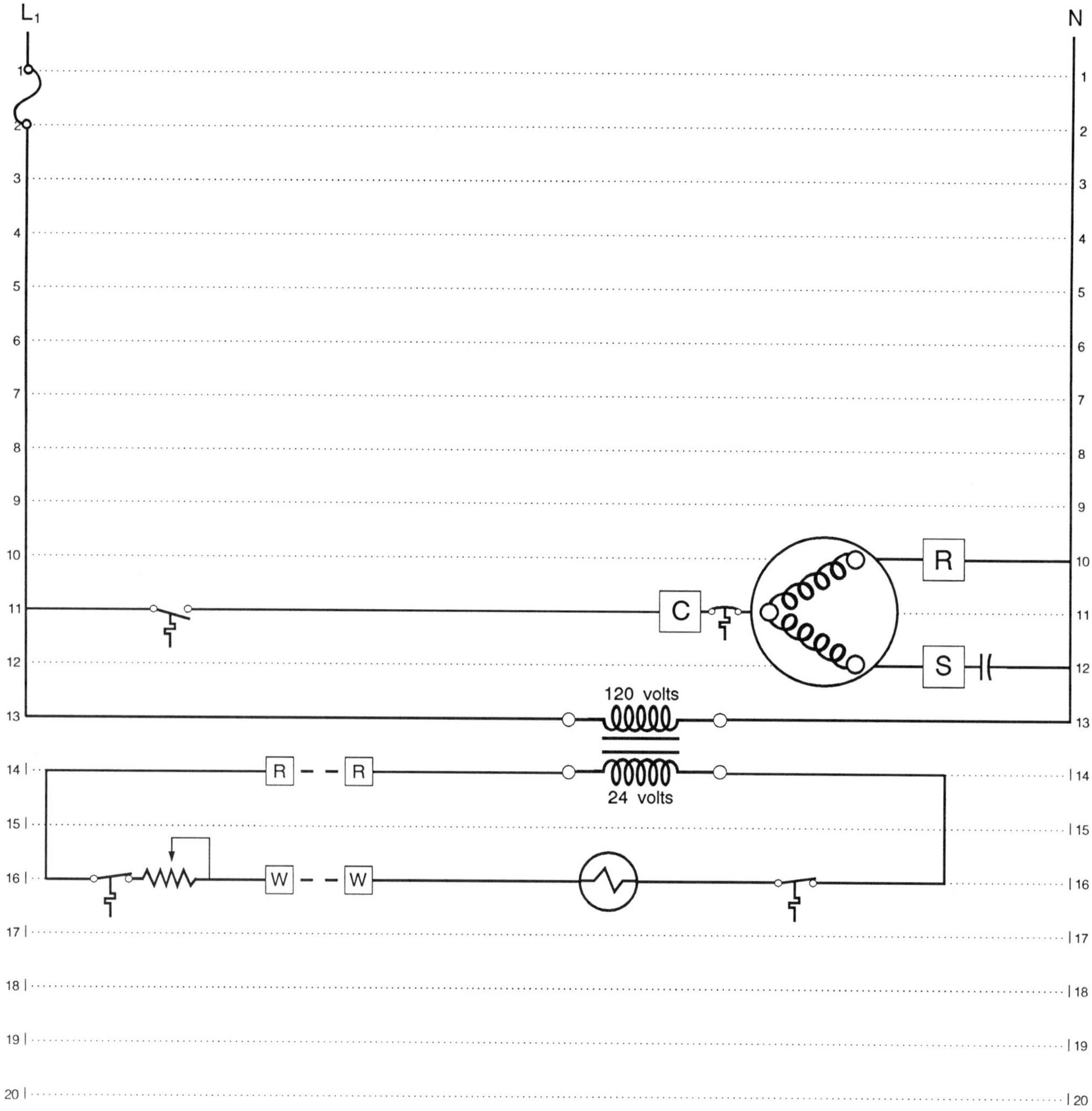
L1
N
R
C
S
120 volts
24 volts
R
R
W
W

EXAMPLE 2
GAS-FIRED, FORCED-AIR FURNACE

The furnace system used in this example is typical of that installed in many homes. The furnace is delivered as a complete system, lacking only the remote thermostat.

1. DEFINE THE FUNCTION OF THE EQUIPMENT

This is a gas-fired, forced-air furnace, operating on line voltage, with automatic, low-voltage controls. Provision is made for uniform air delivery temperatures and a safety control prevents furnace overheating.

2. LIST THE ELECTRICAL COMPONENTS

- The standard home electrical source is 120 V, 60 cycle (60 Hz), single-phase (1ϕ), alternating current (ac).
- A fuse is required by the Safety Code.
- A step-down transformer is required for low (24v) voltage.
- A thermostat with terminals [R] and [W], is required for automatic control.
- A gas valve is required for heat generation.
- A normally-closed thermal limit switch is needed to prevent overheating.
- An internally-protected, permanent split capacitor (PSC) fan motor is needed to drive the fan.
- A normally-open thermal plenum switch is needed for air temperature control.

3. GO STEP-BY-STEP THROUGH THE ELECTRICAL SEQUENCE

- Line voltage (120 vac) must be supplied to the high side of the transformer.
- When the thermostat closes on a temperature fall, the circuit is completed between terminals [R] and [W].
- This induces a 24-V current in the low side of the transformer and energizes the gas valve if the thermal limit switch is closed, as is normal.
- The gas valve can now generate heat and will continue to do so until the plenum overheats at approximately 180 °F. At this temperature, the thermal limit switch opens, and remains open, until the plenum temperature drops to 110 to 120 °F. Meanwhile, the gas valve is off. Until the fan is in operation, the thermal limit switch will cycle the gas valve on and off.
- Fan motor operation is controlled by the normally-open plenum switch, which activates the fan motor when the plenum temperature reaches 140 to 150 °F. Were the fan to operate when the gas valve goes on, the fan would deliver an uncomfortable stream of unheated air on start-up and leave unused heat in the plenum on shutdown. Instead, the fan continues to operate, even with the gas valve off, until the plenum temperature drops below 120 °F and opens the plenum switch.

4. DRAW THE SCHEMATIC

STEP 1. Insert a fuse in L_1 between lines 1 and 2.

STEP 2. Draw a transformer between lines 13 and 14 and connect the 120-V side of the transformer to L_1, and N along line 13.

STEP 3. On line 14, draw terminal [R] and connect it, with low voltage field wiring - - - , to another terminal [R] on the 24-V side of the transformer. Connect the transformer to [R].

STEP 4. On line 16, draw terminal [W] and connect it, with field wiring - - - , to another terminal [W].

STEP 5. Draw a gas valve on line 16 and connect it to terminal [W]. Connect the normally-closed thermal limit switch in series with the gas valve and the 24-V side of the transformer.

STEP 6. On line 16, draw a normally-closed thermostat switch in series with a variable resistor , which is the heat anticipator. Connect these elements to [R] and [W].

STEP 7. Close the thermostat . The gas valve will operate, and generate heat, until...

STEP 8. The thermostat is satisfied or the thermal limit safety opens the circuit to the gas valve .

STEP 9. On line 11, draw the normally-closed internal protector between terminal [C] and the junction of the start and run windings [R]. Connect the run winding to N along line 10. Insert a capacitor [S] between the start winding and terminal [S]. Connect [S] to N along line 12.

STEP 10. On line 11, connect a normally-open plenum fan switch to L_1, and terminal [C].

STEP 11. Close the thermostat to start the gas valve . Wait one minute, to allow heat to reach the plenum fan switch . When the heat-operated plenum switch closes, the system is in normal operation.

STEP 12. Open the thermostat. The plenum fan switch will remain closed, and the fan will continue to operate until the plenum cools down. The plenum fan switch will then open and the system will be on standby.

The tear-out
schematic layout sheet
can be found on page 227.

STEP 1

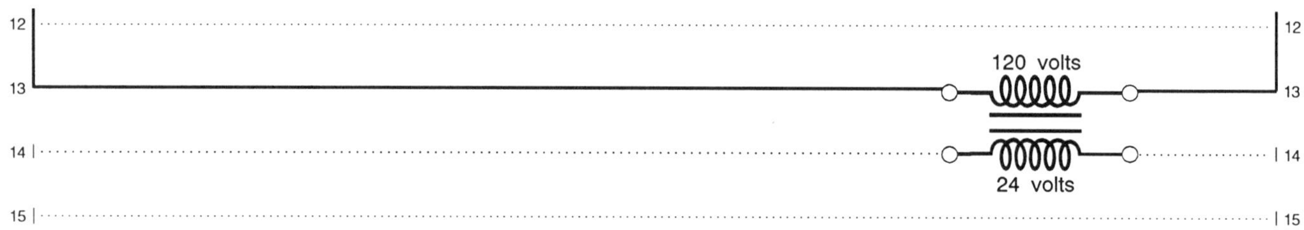

STEP 2

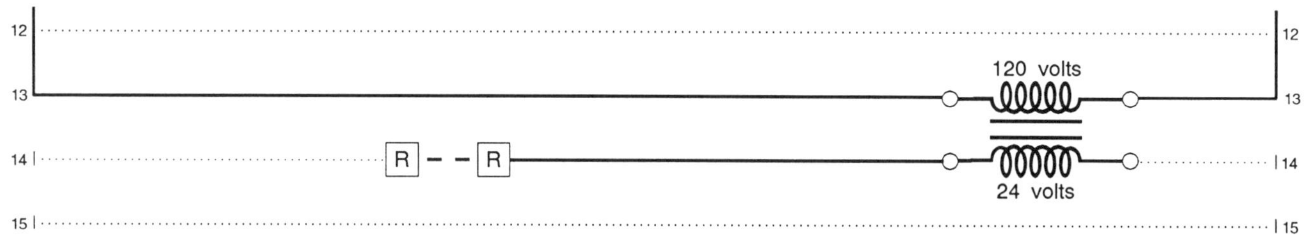

STEP 3

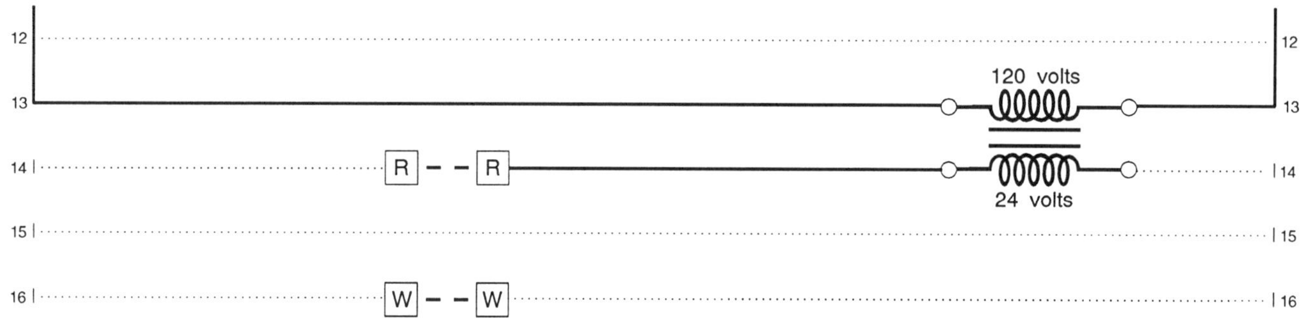

STEP 4

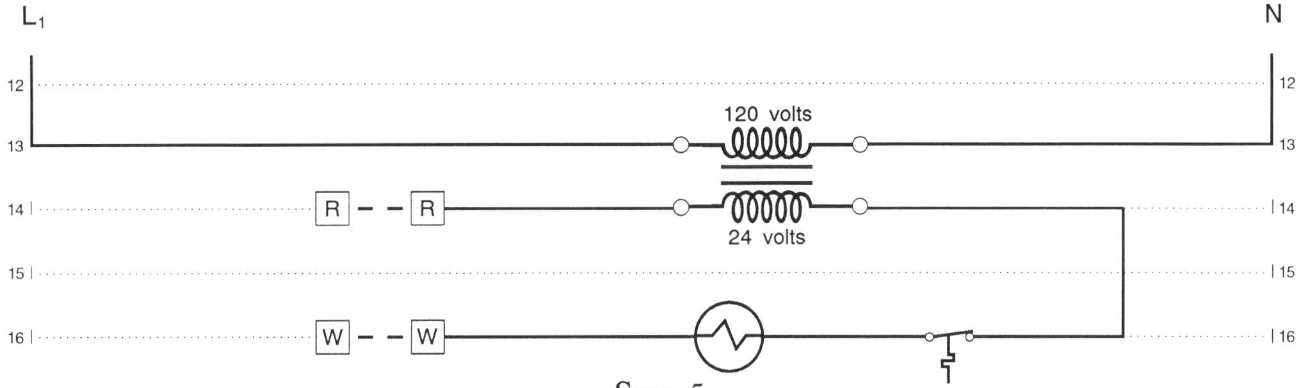

Step 5

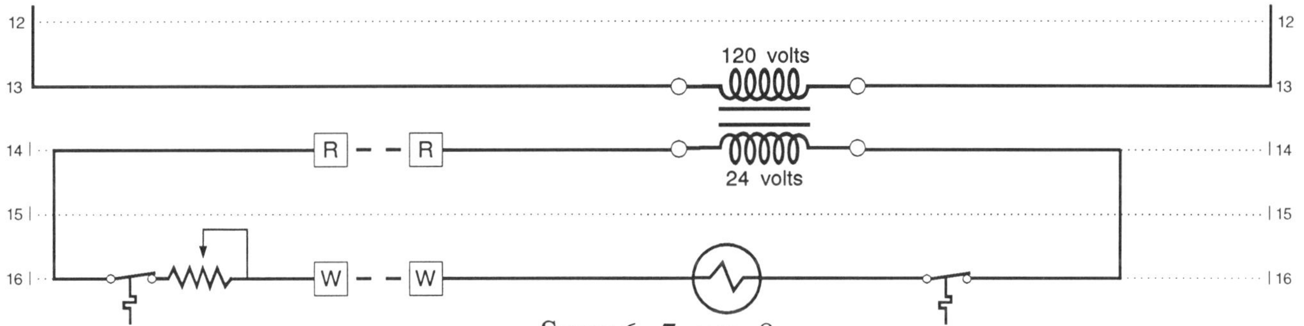

Steps 6, 7 and 8

Step 9

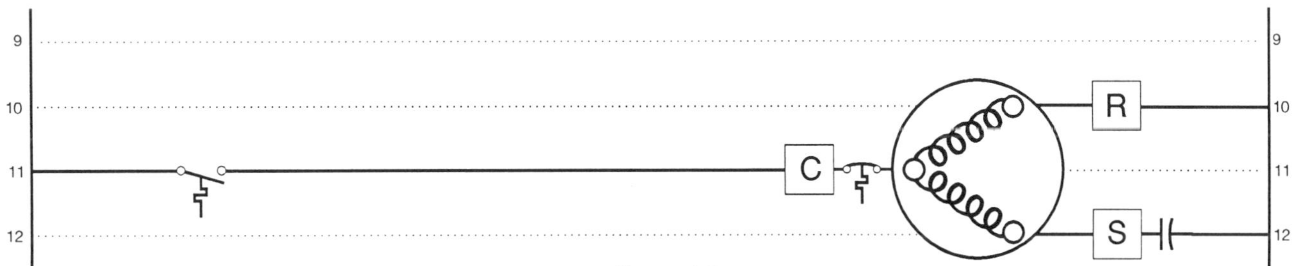

Step 10

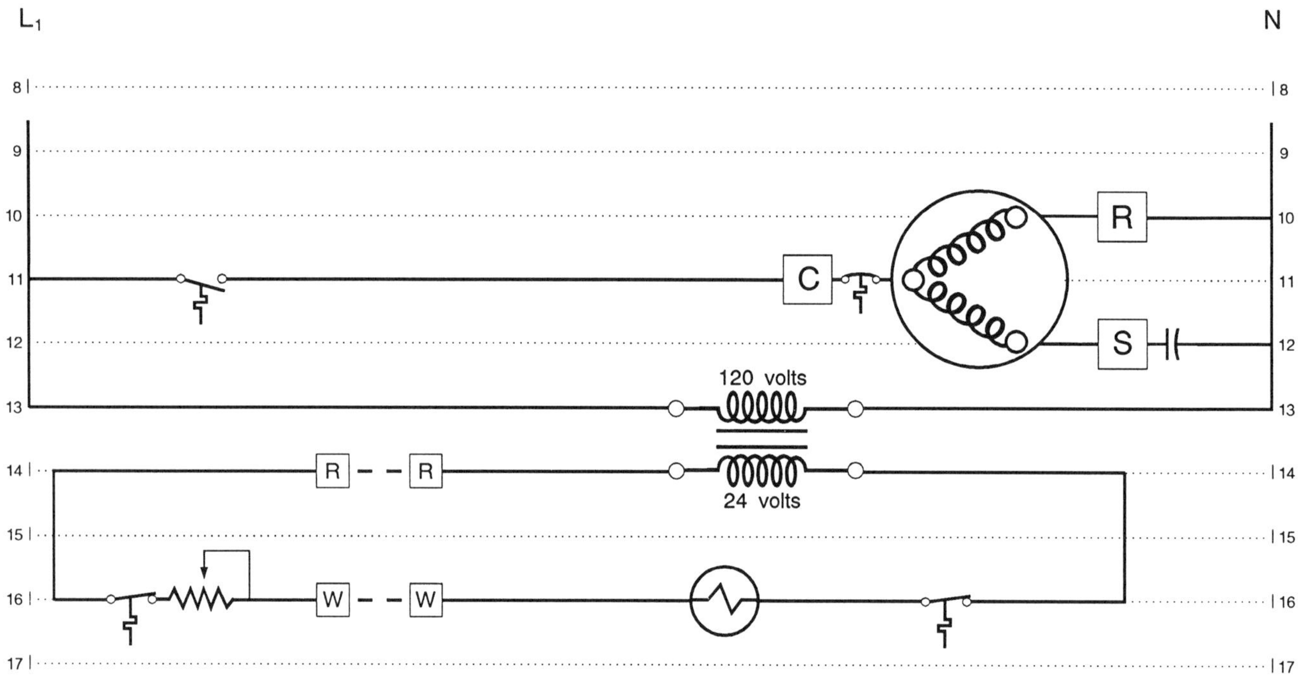
L1
N
8
9
10
11
12
13
14
15
16
17
C
R
S
120 volts
24 volts
R
R
W
W

QUESTIONS FOR EXAMPLE 2

What voltages would you expect to find at the following contacts and loads? Read the question carefully before answering.

Q-1, The plenum fan switch, line 11, with the burner on and the fan running?
A-1, ______V

Q-2, The plenum fan switch, line 11, with the burner off and the fan off?
A-2, ______V

Q-3, Limit switch, line 16, with the burner on?
A-3, ______V

Q-4, Limit switch, line 16, with the burner off?
A-4, ______V

Q-5, Heating thermostat contacts, [R] to [W], with the heat on?
A-5, ______V

Q-6, Heating thermostat contacts, [R] to [W], with the heat off?
A-6, ______V

Q-7, Gas valve, line 16, with the heat on?
A-7, ______V

Q-8, Gas valve, line 16, with the heat off?
A-8, ______V

Q-9, Terminals [R] to [C] on the fan motor with the motor running?
A-9, ______V

Q-10, Motor terminals [R] to [S] with the motor running?
A-10, ______V

Answer the following questions. List, for each problem, all of the electrical items that could be checked for defect.

Q-1, The indoor fan motor did not come on when the burner was on (there is a flame).
A-1, ______________________________

Q-2, Thermostat contacts [R] to [W] closed, but the gas valve did not open (no flame).
A-2, ______________________________

Q-3, The fan motor is running but the gas burner is off (no flame).
A-3, ______________________________

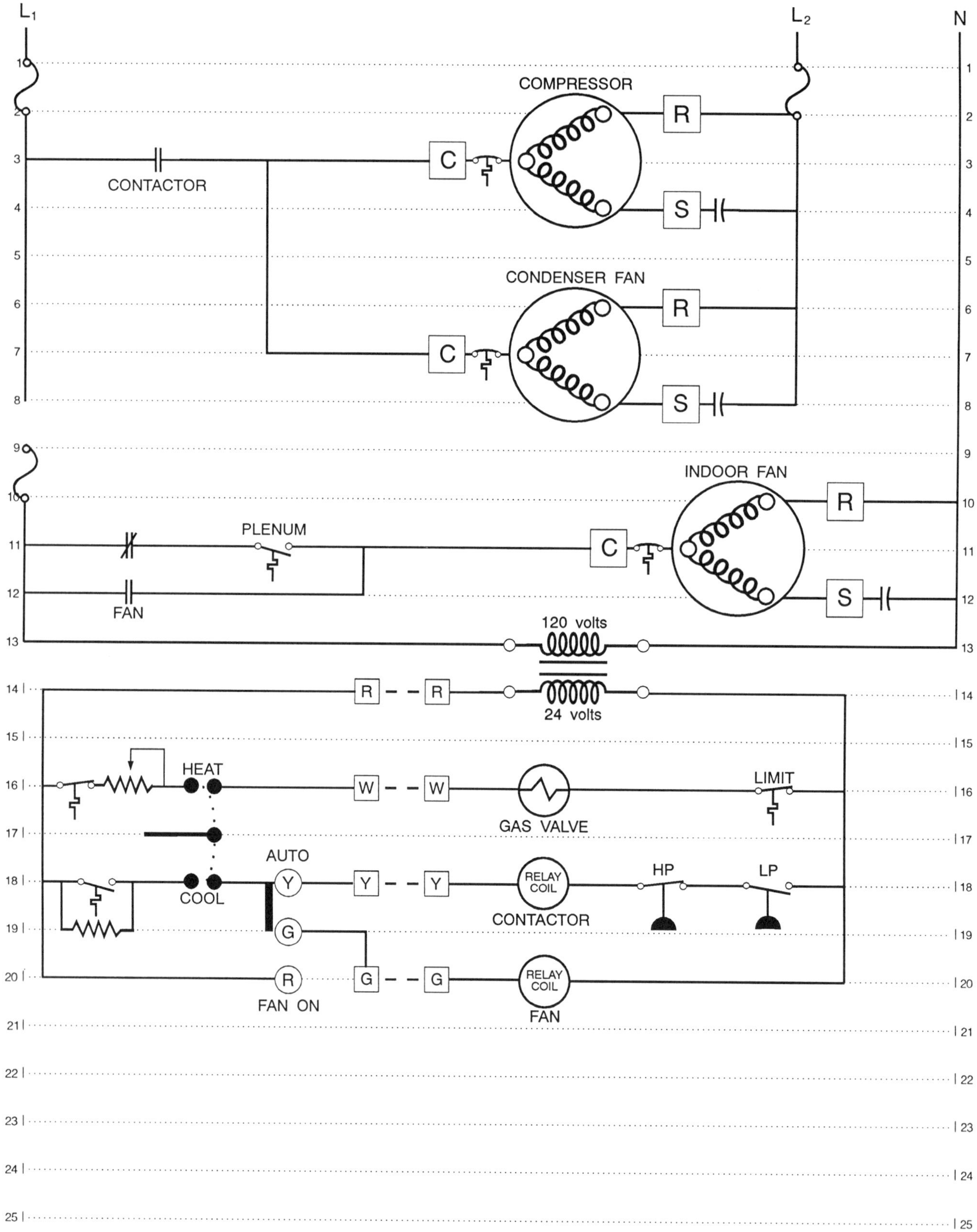

L1
L2
N
COMPRESSOR
CONTACTOR
CONDENSER FAN
INDOOR FAN
PLENUM
FAN
120 volts
24 volts
HEAT
LIMIT
GAS VALVE
AUTO
COOL
HP
LP
RELAY COIL
CONTACTOR
FAN ON
RELAY COIL
FAN

EXAMPLE 3
GAS-FIRED, FORCED-AIR FURNACE WITH AUTOMATIC AIR CONDITIONING

All of the controls shown in this heating/cooling combination are prewired at the factory, with the exception of the thermostat, which is furnished by the installer. For simplicity's sake, field wiring between the furnace and the outside-mounted compressor/condenser unit will not be indicated.

1. DEFINE THE FUNCTION OF THE EQUIPMENT

This is a gas-fired, forced air furnace and air conditioner, operating on line voltage, with automatic, low voltage, controls.

The heating side of this combination is designed to provide uniform air delivery temperatures. It has a safety control to prevent overheating. On the cooling side, the compressor is protected by high pressure and low pressure cut-outs.

A forced-air fan operates when either heating or cooling is called for and, as an added feature, can be operated continuously while either system is on standby.

2. LIST THE ELECTRICAL COMPONENTS

- The standard home electrical source is 120 V - 240 V, 60 cycle (60 HZ), single-phase (1ϕ), alternating current (ac).
- Two fuses are required by the Safety Code.
- A step-down transformer is required for low (24 V) voltage.
- A heating /cooling thermostat with terminals R (common), W (heat), Y (cooling), and G (fan), is required for automatic control.
- A gas valve is required for heat generation.
- A normally-closed thermal limit switch is needed to prevent overheating.
- An internally-protected, permanent split capacitor fan motor (PSC) is needed to drive the forced-air fan.
- A normally-open, thermal plenum switch is needed for air temperature control.
- High pressure and low pressure cut-out switches are needed to protect the compressor.
- A contactor relay with a coil and a set of contacts is needed to start the compressor and condenser fan motors.
- A fan relay with a coil, and one set of normally-closed and one set of normally- open contacts, is needed to permit continuous fan operation.
- An internally-protected, permanent split capacitor compressor motor and condenser fan motor are both needed for cooling.

3. Go Step-by-Step Through the Electrical Sequence

HEATING

- Line voltage (120 V) must be supplied to the primary side of the transformer.
- When heating is desired, the manual mode switch on the thermostat must be set on HEAT and the manual fan control switch is usually on AUTO. When the heating thermostat closes on a temperature fall, a circuit is completed between terminals R and W. This completes a circuit through the low (24 V) side of the transformer to the gas valve and normally-closed thermal limit switch, to start the heating cycle. When the temperature in the plenum reaches 140° to 150 °F, the normally-open fan switch closes to complete a line voltage circuit through the normally-closed side of the indoor fan relay to the indoor fan motor.

COOLING

- When cooling is desired, the manual mode switch on the thermostat must be set on COOL. The manual fan control switch is usually on AUTO. When the cooling thermostat closes on a temperature rise, a circuit is completed between terminals R and Y. This completes a circuit through the low side of the transformer to the contactor relay coil and through the normally-closed high pressure and low pressure cut-out switches. The contactor coil closes the contacts and completes parallel circuits to the start winding of the compressor and condenser fan motors. At the same time, another circuit is completed from R to G through the indoor fan relay coil. When the coil is energized, it opens the normally- closed set of contacts and closes the normally-open set of contacts. As a result, the normally-open fan switch is bypassed and a circuit is completed to the common terminal of the indoor fan motor.
- When continuous fan operation is required, the manual fan switch is moved to On. This completes a circuit from R to G and energizes the indoor fan relay coil. With the plenum switch bypassed through the indoor fan relay, the fan operates continuously, even though neither heating nor cooling is called for. The fan will continue to operate when either heating or cooling are desired.

4. DRAW THE SCHEMATIC

STEP 1. Insert a fuse in line L_1 between lines 1 and 2 and between lines 9 and 10.

STEP 2. Insert a fuse in line L_2 between lines 1 and 2.

STEP 3. Draw a step-down transformer between lines 13 and 14 and connect the 120-V side to L_1 and N along line 13.

STEP 4. Draw four pairs of field wired _ _ _ terminals, [R] [R] on line 14, [W] [W] on line 16, [Y] [Y] on line 18, and [G] [G] on line 20.

STEP 5. On line 14, connect terminal [R] to the 24-V side of the transformer.

STEP 6. On line 16, draw a normally-closed, temperature-actuated switch (the heating thermostat) in series with a variable resistor (the heat anticipator). Join the heat anticipator to a pair of open contact buttons ● ● and connect the buttons to terminal [W] on the thermostat. Continuing on line 16, connect a gas valve between terminal [W] and a normally-closed, temperature actuated limit switch. Link the limit switch to the return side of the transformer on line 14.

STEP 7. On line 18, draw a normally-open, temperature-actuated switch (the cooling thermostat) and connect it to a pair of open contact buttons. Connect the buttons to terminal [Y] on the thermostat. Continuing on line 18, connect the contactor relay coil (RELAY COIL) to its terminal [Y] and to a pair of normally-closed high and normally-open low pressure switches. Link the pressure switches to the return side of the transformer on line 14.

STEP 8. On line 19, draw a variable resistor (the cooling anticipator) and wire it parallel with the cooling thermostat by connecting it to line 18.

STEP 9. Close the gap between the two contact buttons ● ● on line 16 with a third button ● attached to an arm that pivots from line 17. This arm is the heat/cool selector switch. It can close the gap between the two buttons on line 16 to complete the heating circuit and isolate the cooling circuit. Or, it can close the gap between the two buttons on line 18 to complete the cooling circuit and isolate the heating circuit.

STEP 10. On line 20, connect a fan relay coil (RELAY COIL) between its terminal [G] and the return side of the transformer.

STEP 11. Draw the internal fan contacts by inserting terminal (Y) on line 18, terminal (G) on line 19, and terminal (R) on line 20. Connect terminal (G) to terminal [G] on line 20. Connect terminal (G) and the two thermostat switches to terminal [R] on line 14.

Connect terminal (Y) to terminal (G) with a sliding bar |. With the bar in the UP or AUTO position, joining (Y) and (G), the fan relay is controlled by the thermostat. With the bar in the DOWN, or FAN ON position, joining (G) and (R), the fan runs constantly, regardless of the thermostat action. This completes the low- voltage circuits.

The tear-out
schematic layout sheet
can be found on page 229.

STEP 12. Draw two internally-protected PSC motors with their common C terminals on lines 3 and 7. Connect the run R and start S terminals of the compressor and condenser motors to L_2.

STEP 13. Draw one internally-protected PSC fan motor with the common C terminal on line 11. Connect the run R and start S terminals to N.

STEP 14. Draw the contactor contacts for the compressor motor on line 3 and connect them to L_1 and terminal C of the compressor.

STEP 15. On line 7, connect terminal C of the condenser fan to the compressor on line 3.

STEP 16. On line 11, draw a pair of normally-closed fan contacts and a normally-open plenum switch and join them to the C terminal of the indoor fan motor and L_1.

STEP 17. On line 12, draw in the normally-open set of indoor fan relay contacts. Wire them from L_1 to the fan motor, bypassing the plenum fan switch.

STEP 18. Recheck the low-voltage circuits to determine how the compressor and compressor fan are started simultaneously, how the internal fan operates on HEAT and on COOL, and how the internal fan motor is operated when neither heating nor cooling is called for.

STEP 1

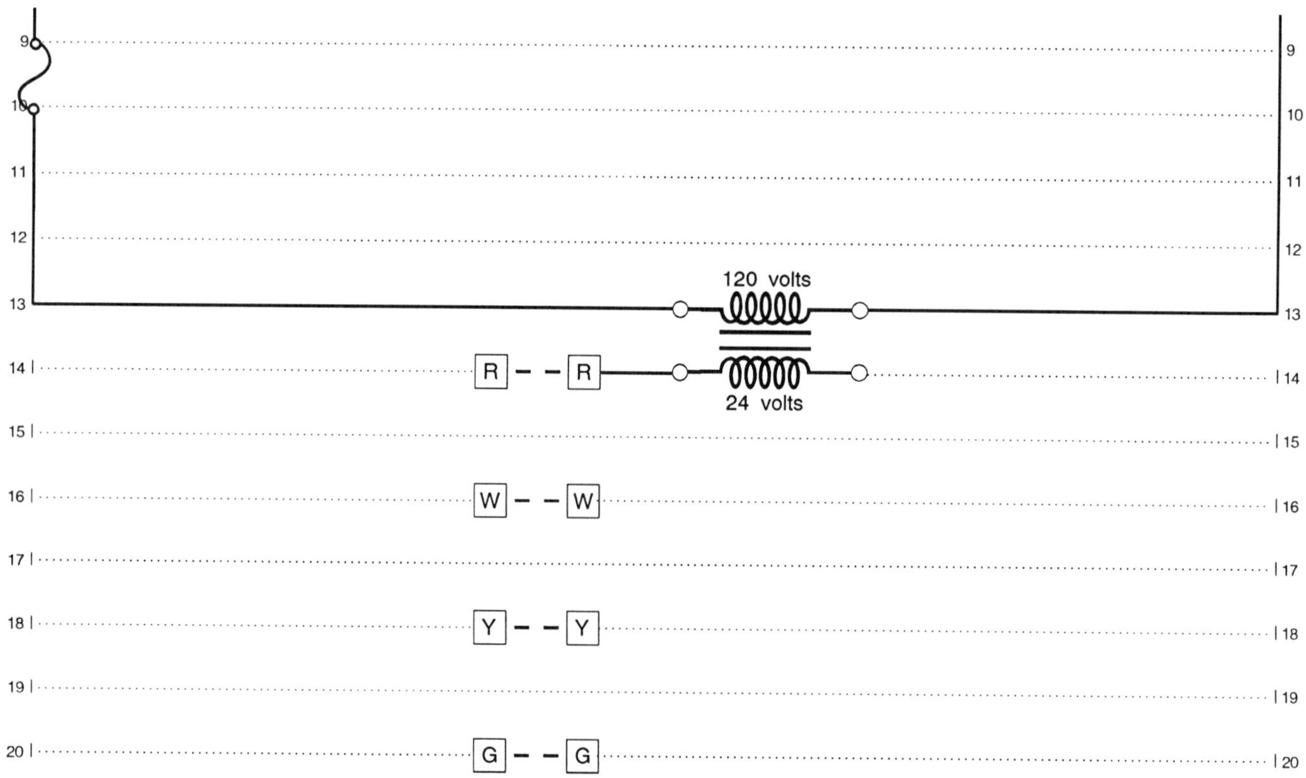

STEPS 2, 3, 4 AND 5

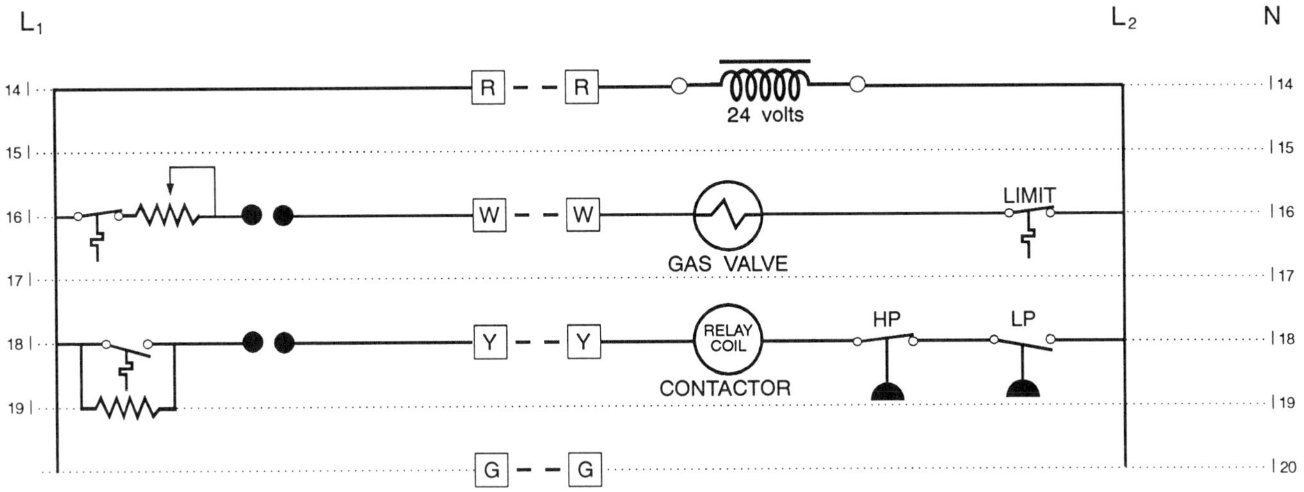

STEPS 6, 7 AND 8

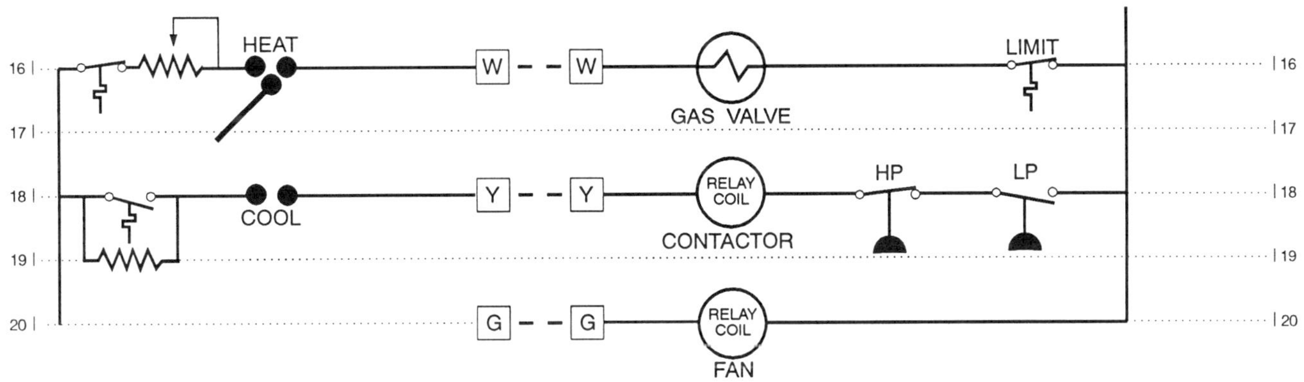

STEPS 9 AND 10

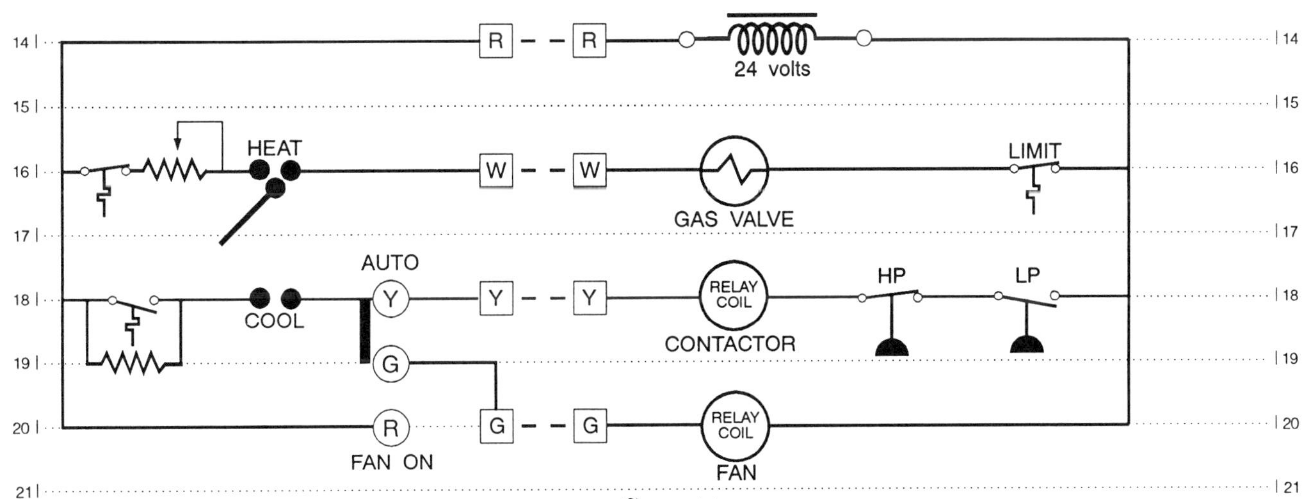

STEP 11

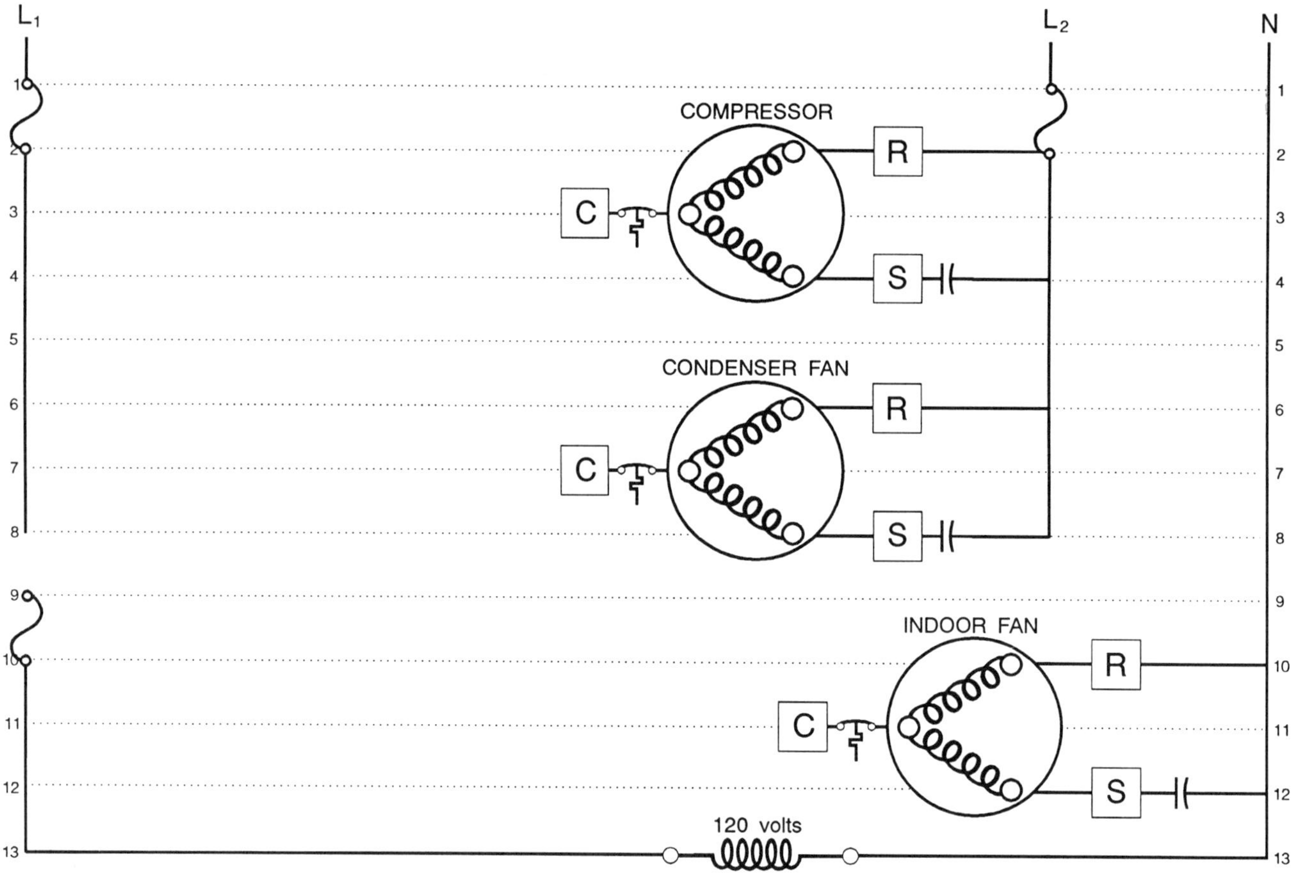

STEPS 12 AND 13

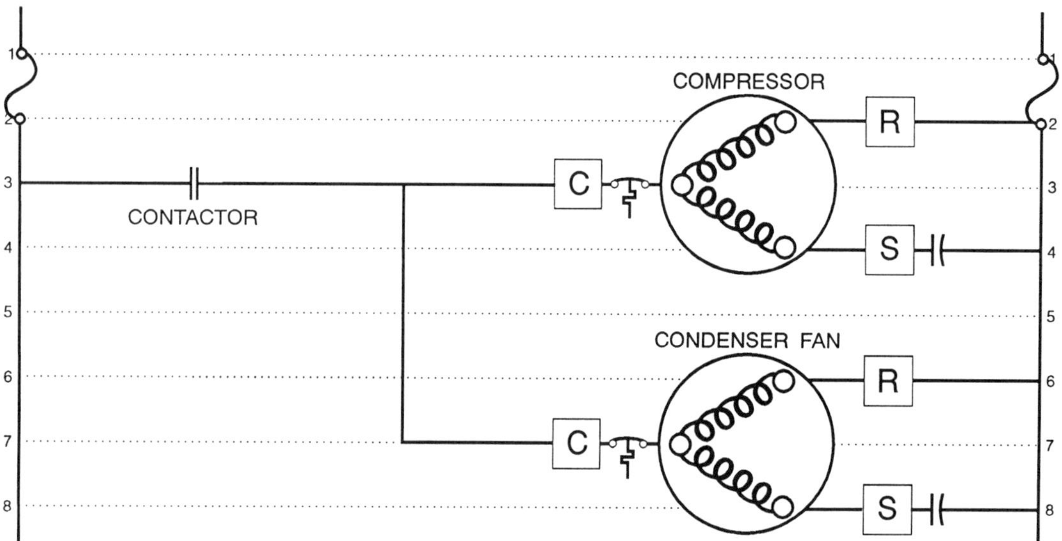

STEPS 14 AND 15

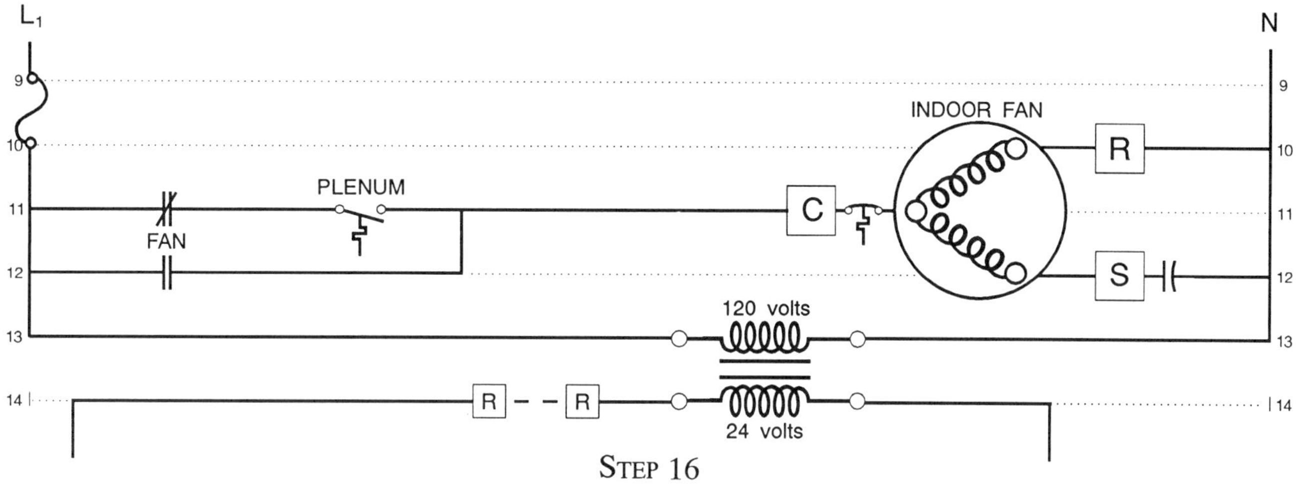

STEP 16

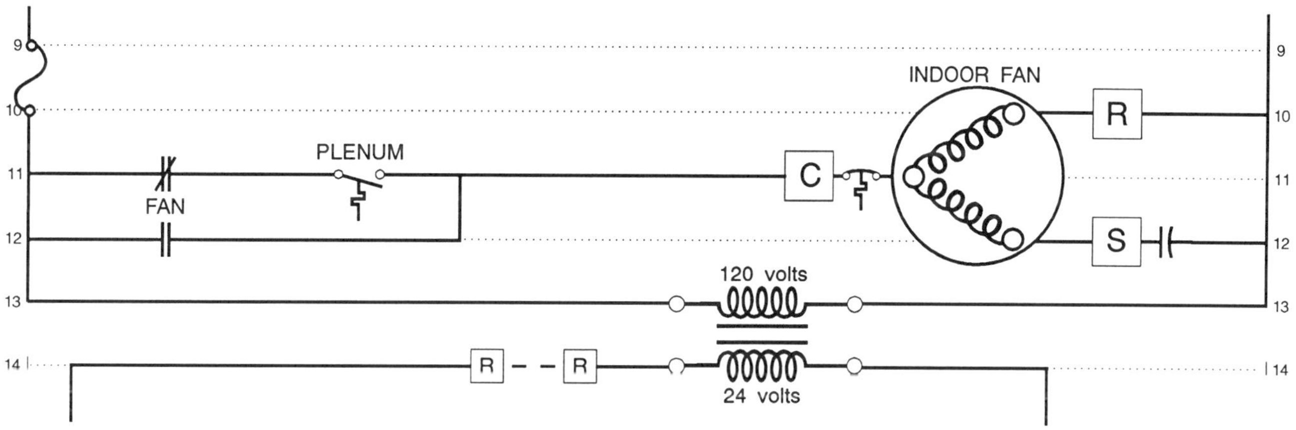

STEP 17

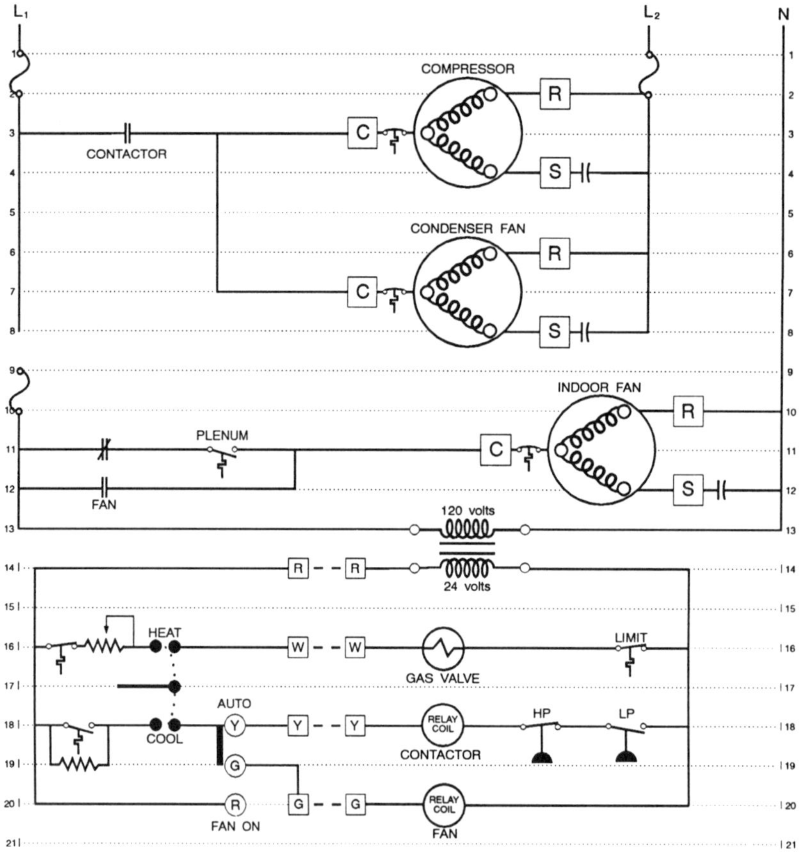

QUESTIONS FOR EXAMPLE 3

What voltage would you expect to find at the following contacts and loads? Read the question carefully before answering.

Q-1, The plenum switch, line 11, with the fan motor running and the heat on?

A-1, _____V

Q-2, The plenum switch, line 11, with the fan motor running and the system operating on cooling?

A-2, _____V

Q-3, The low pressure switch, line 18, with the system operating on cooling?

A-3, _____V

Q-4, The high pressure switch, line 18, with the cooling system in the OFF position?

A-4, _____V

Q-5, The contactor relay coil, line 18, with the system operating on cooling?

A-5, _____V

Q-6, The contactor relay coil with the thermostat in the OFF position?

A-6, _____V

Q-7, Terminals R to G, line 20, with system switch in the OFF position?

A-7, _____V

Q-8, Terminals [R] to [G], line 20, with the heating thermostat contacts closed?
A-8, _____V

Q-9, Terminals [R] to [G], line 20, with the cooling thermostat contacts closed?
A-9, _____V

Q-10, Transformer secondary with fuse L_1 open?
A-10, _____V

Q-11, Terminals [R] to [W], lines 14 and 16, with the heating thermostat contacts open?
A-11, _____V

Q-12, Terminals [R] to [Y], lines 14 and 18, with the cooling thermostat contacts open?
A-12, _____V

Q-13, Contactor contacts, line 3, with the heating thermostat contacts open?
A-13, _____V

Q-14, Contactor contacts, line 3, with the cooling contacts closed?
A-14, _____V

Q-15, Terminals [R] to [C], lines 2 and 3, with the compressor running?
A-15, _____V

Q-16, Condenser fan motor terminals [R] to [S], lines 6 and 8, with the motor running?
A-16, _____V

Q-17, Contactor relay coil, line 18, with the thermostat closed in the cooling position?
A-17, _____V

Q-18, Fan contacts, line 12, with the system operating on heating?
A-18, _____V

Q-19, Gas valve, line 16, with the limit switch open and the system on heating?
A-19, _____V

Q-20, Gas valve, line 16, with the thermostat in the cooling position?
A-20, _____V

List, for each problem, all of the electrical items that could be checked for defects.

Q-1, The indoor blower fan motor is running, and the thermostat heating contacts are open.
A-1, ______________________________

Q-2, The compressor is operating but the condenser fan motor is off.
A-2, ______________________________

Q-3, The cooling thermostat contacts are closed, the indoor fan motor is running, but the condenser motor and compressor are off.
A-3, ______________________________

Q-4, The thermostat is in the cooling position, the compressor and condenser fan motors are running, but the indoor fan motor is off.
A-4, ______________________________

Q-5, The heating side of the system runs perfectly, but the cooling side does not work at all.
A-5, ______________________________

EXAMPLE 4 OIL-FIRED, FORCED-AIR FURNACE

This oil heating system uses the same air temperature controls as a gas-fired system. What is unique is a burner primary control that incorporates a number of safety features to shut down the oil delivery system in the event of ignition failure or termination. The primary control is an encapsulated unit that cannot be repaired. However, its components will be reviewed in order to understand the operation of the safety features.

1. DEFINE THE FUNCTION OF THE EQUIPMENT

This is an oil-fired, forced-air furnace requiring a low-voltage thermostat and having air delivery temperature controls and burner ignition controls as safety features.

2. LIST THE ELECTRICAL COMPONENTS

The standard home electrical source is 120 V, 60 cycle (60 Hz), single-phase (1ϕ), alternating current (ac).

- A fuse is required by the Safety Code.
- An internally-protected, permanent split capacitor fan motor is needed to drive the forced-air fan.
- A normally-open plenum switch is needed for air temperature control.
- A normally-closed thermal limit switch is needed to prevent overheating.
- A burner motor relay coil is needed to control the burner motor that both pumps the oil and supplies combustion air.
- A 10,000-V ignition transformer and electrodes are needed to ignite the oil.
- A cad cell flame detector is needed to ensure shutdown on flame-out.
- An oil burner primary control is needed to supply low, 24 V and for ignition control.
- A thermostat with terminals R (common) and W (heat) is required for automatic control.

3. GO STEP-BY-STEP THROUGH THE ELECTRICAL SEQUENCE

- 120-V line voltage must be supplied to the plenum switch, the thermal limit switch and the oil burner primary control terminals L_1 and N.
- When the thermostat closes on a temperature fall, a circuit is established between R and W in the thermostat and between T and T in the primary control. The circuit is completed through the low-voltage side of the primary control transformer, the normally-closed flame relay contacts, the safety heater, the normally-closed safety switch, and the burner motor relay coil.
- When energized, the burner motor relay coil closes two sets of normally-open contacts. One set completes a low-voltage circuit that parallels the flame relay contacts and the safety switch heater circuit. The other set of contacts completes a line-voltage circuit, through the normally-closed thermal limit switch, to the burner motor. and the ignition transformer.
- The ignition transformer creates a 10,000-V arc between two electrodes to ignite the oil being pumped by the burner motor.
- If the oil is ignited, a cad cell flame detector will detect the flame and complete a low-voltage circuit through the flame relay coil. When energized, the flame relay coil will open the flame relay contacts and interrupt the circuit to the safety heater.

- If the oil fails to ignite, or ignition ceases, the flame relay contacts —╫— will stay closed and the safety heater —ᐱᐱᐱ— will remain in the circuit. After several seconds, the heater will warp open the safety switch —╫— and interrupt the circuit to the burner motor relay (RELAY COIL), thereby opening all of the control circuits.
- The oil burner primary control will not recycle until the safety switch heater has cooled and the safety switch has been closed by pressing a manual reset button.

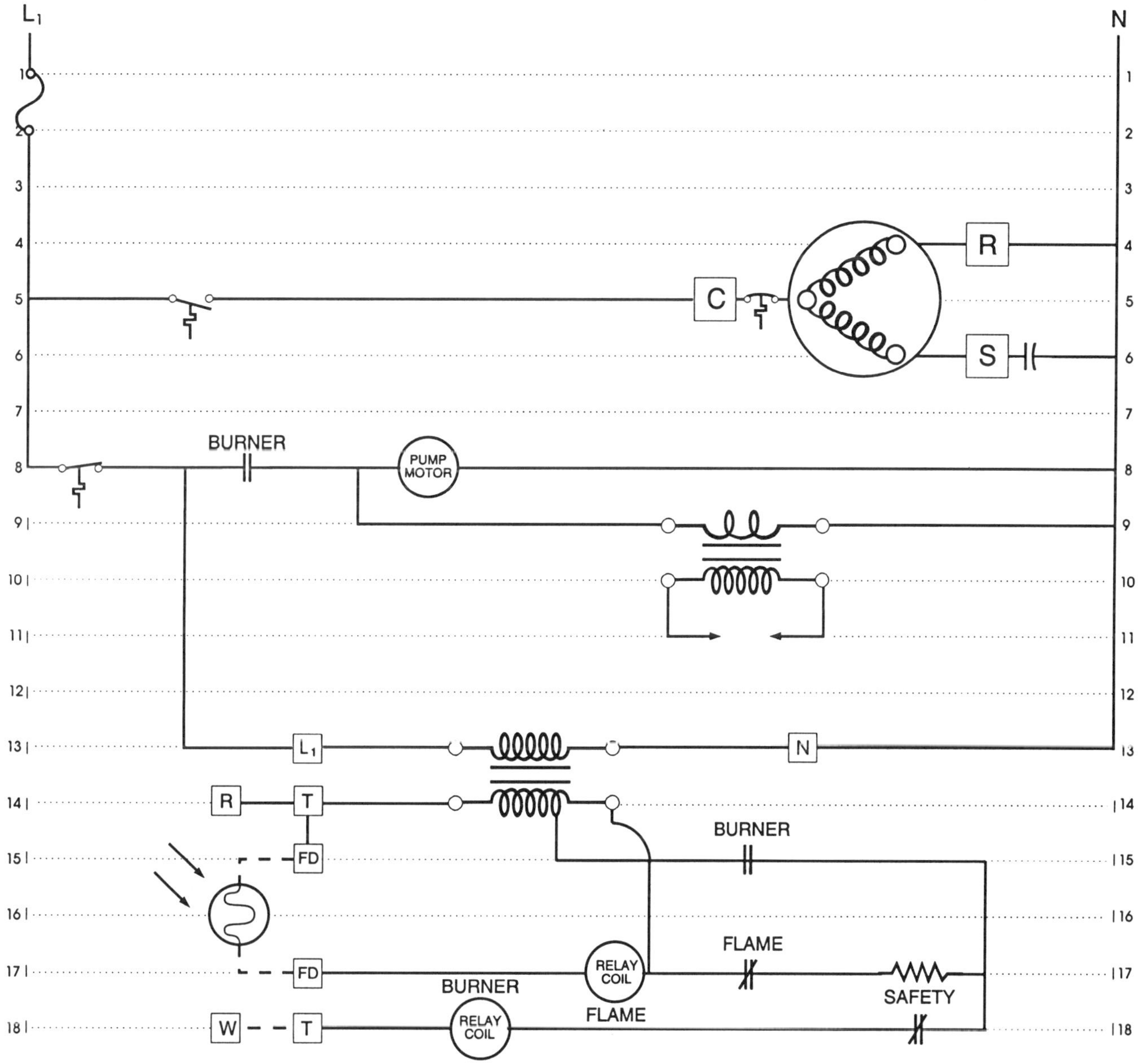

4. DRAW THE SCHEMATIC

STEP 1. Insert fuse in L_1 between lines 1 and 2.

STEP 2. Draw an internal protector for the indoor fan motor on line 5. Connect the run winding to N along line 4 and the start winding and run capacitor —|(— to N along line 6. Insert a normally-open plenum switch on line 5 and connect it to L_1 and the internal protector.

STEP 3. Draw a burner motor. on line 8. Draw a normally-closed thermal limit switch and a set of normally-open burner motor relay contacts —||—, between L_1 and the burner motor. Connect the return side of the burner motor to N.

STEP 4. Draw a step-up ignition transformer on lines 9 and 10. Draw two-electrodes on line 11 and connect them to the high side of the transformer. Connect one leg of the low side to N along line 9 and the other leg to line 8 at a point between the relay contacts and the burner motor.

STEP 5. Draw a step-down transformer on lines 13 and 14 . Connect the 120-V side to terminals L_1 and N of the burner primary control on line 12, and connect N to the line voltage neutral. Connect L_1 between the thermal limit switch and burner motor contacts —||— on line 8. This low voltage transformer and all of the low voltage circuits are integral parts of the burner primary control.

Step 6. Draw thermostat terminals R and W on lines 14 and 18. Draw the following terminals as part of the primary control: T on lines 14 and 18, FD on lines 15 and 17.

Step 7. Field wire – – – R to T on line 14, W to T on line 18. Connect T on line 14 to FD on line 15. Draw a flame detector on line 16 and field wire – – – it to FD and FD. Connect the 24-V side of the transformer to T on line 14.

STEP 8. On line 18, connect a burner motor relay coil and a normally-closed safety switch —||/— to T. On line 17 connect a safety switch heater —/\/\/\—, a normally-closed flame relay switch —||/—, and the flame relay coil to FD.

STEP 9. Draw a second set of burner motor contacts —||— on line 15. Connect them to the transformer tap and to the safety switch and the safety switch heater —/\/\/\—.

The tear-out
schematic layout sheet
can be found on page 231.

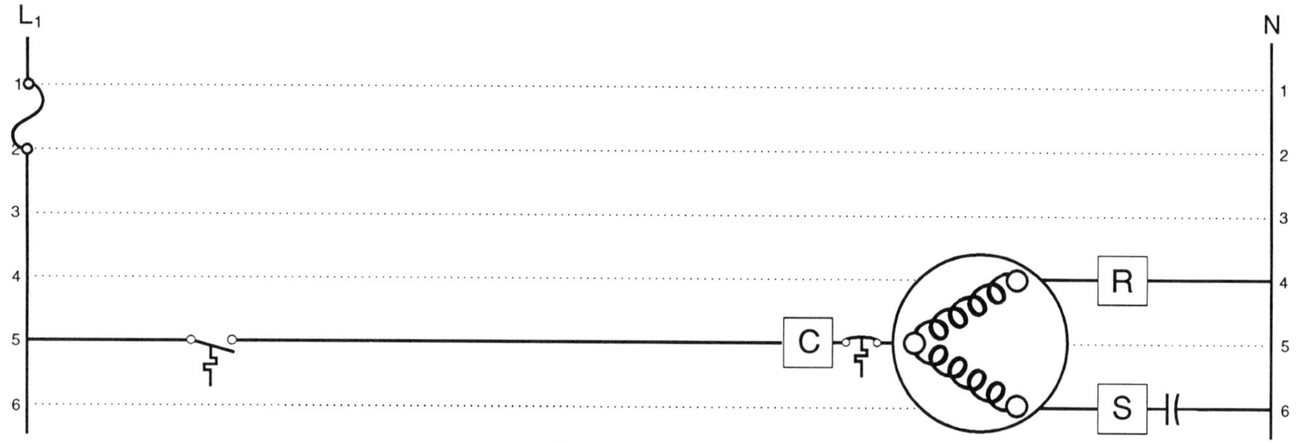

STEPS 1 AND 2

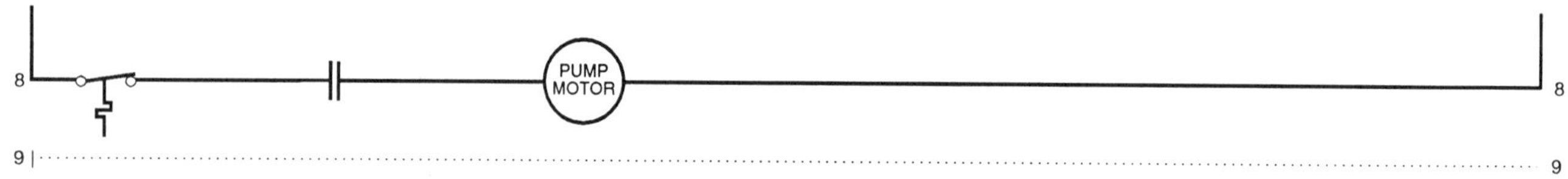

STEP 3

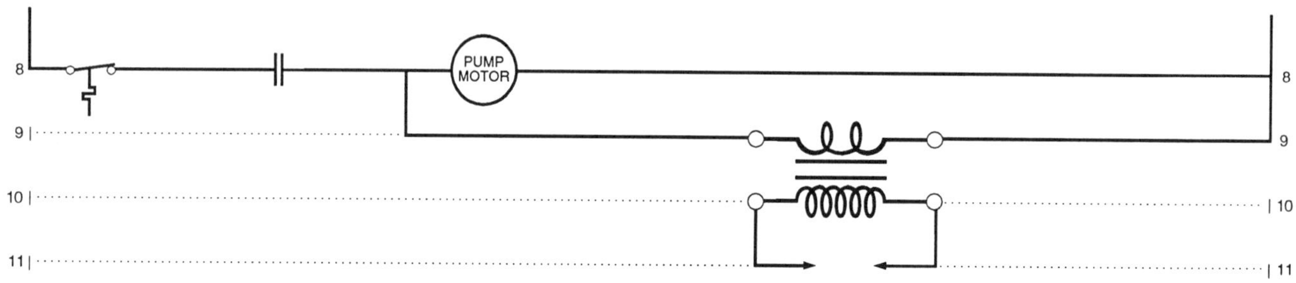

STEP 4

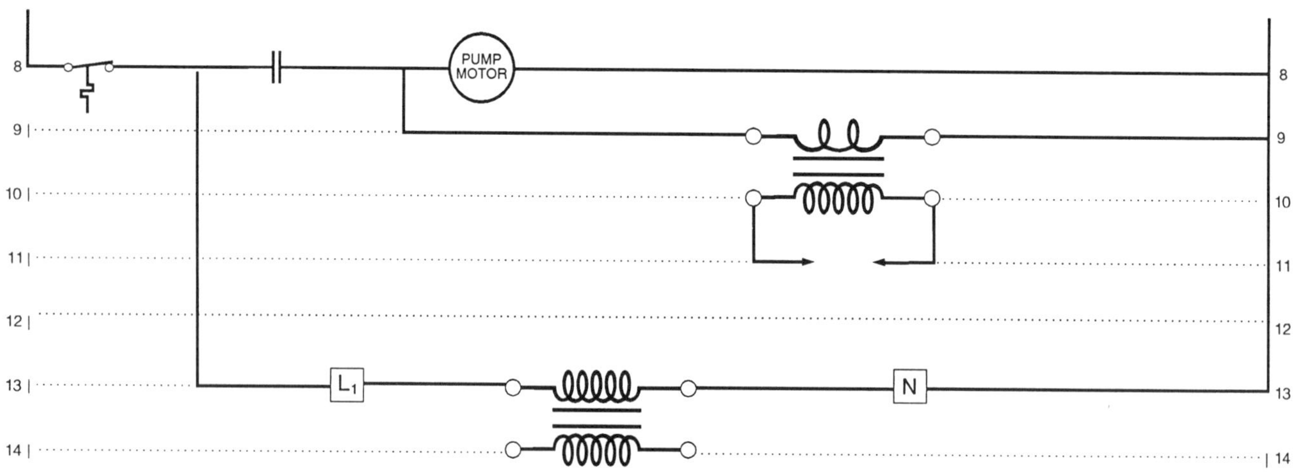

STEP 5

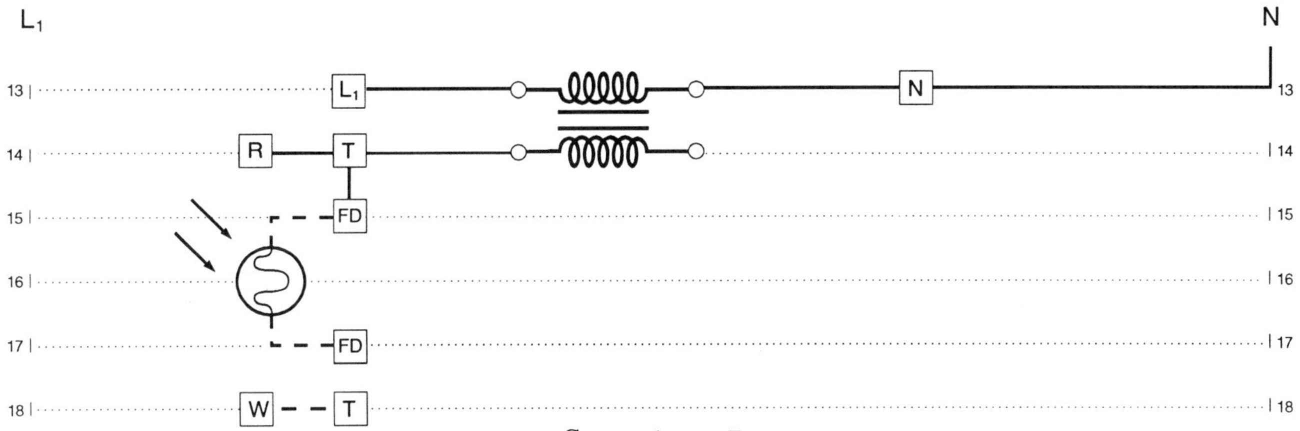

STEPS 6 AND 7

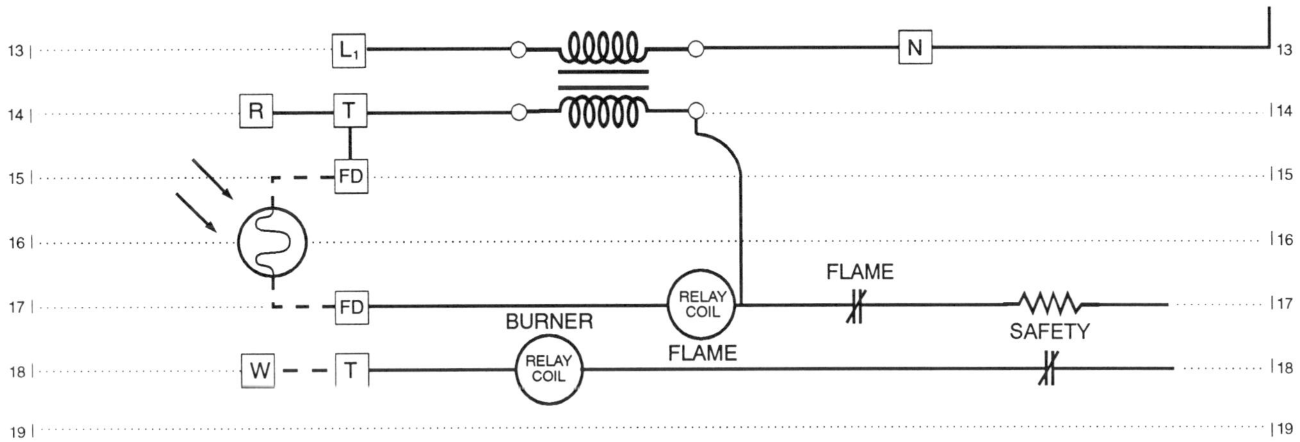

STEP 8

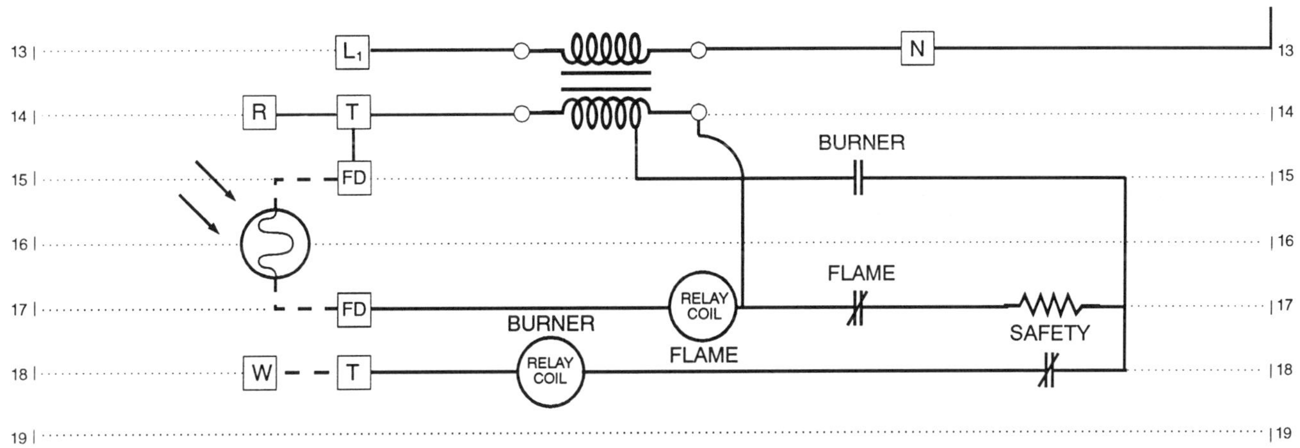

STEP 9

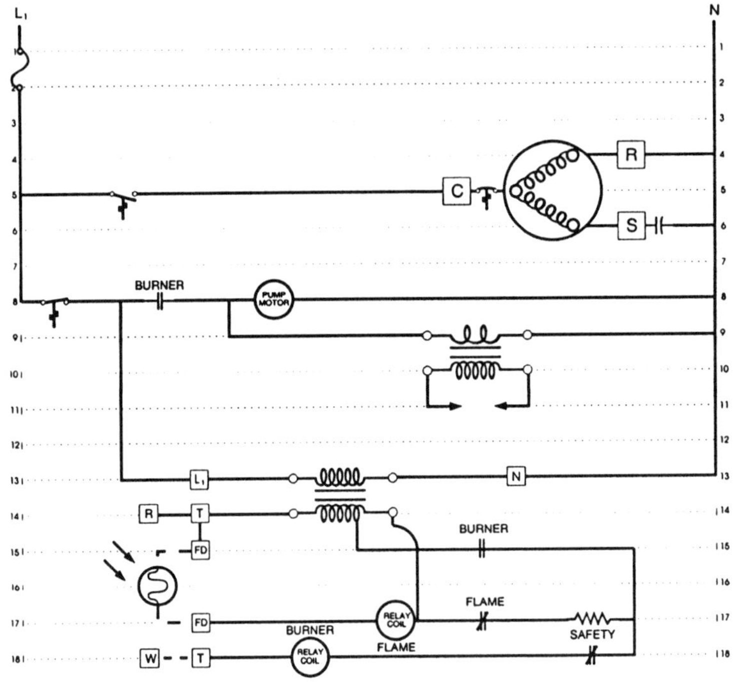

QUESTIONS FOR EXAMPLE 4

What voltages would you expect to find at the following contacts and loads? Read the question carefully before answering.

Q-1, The plenum switch, line 5, with the fan motor running?

A-1, ____V

Q-2, The limit switch, line 8, with the burner on? (Flame)

A-2, ____V

Q-3, The limit switch with the thermostat in the OFF position?

A-3, ____V

Q-4, Ignition transformer secondary, burner on? (Flame)

A-4, ____V

Q-5, Ignition transformer primary, line 9, with burner on? (Flame)

A-5, ____V

Q-6, Ignition transformer secondary with the burner off? (No flame)

A-6, ____V

Q-7, Ignition transformer secondary with the thermostat in the OFF position?

A-7, ____V

Q-8, Pump motor, line 8, with the burner relay contacts open?

A-8, ____V

Q-9, Pump motor with the burner relay contacts closed?

A-9, ____V

Q-10, Ignition transformer secondary with the limit, line 8, open?
A-10, ____V

Q-11, 24-volt transformer secondary, line 14, with the limit open?
A-11, ____V

Q-12, 24-volt transformer primary, line 13, with the limit open?
A-12, ____V

Q-13, Terminals [R] to [W], line 14 and 18, with the safety switch, line 18, open?
A-13, ____V

Q-14, Flame relay coil, line 17, with no burner flame and the heating thermostat contacts closed?
A-14, ____V

Q-15, Terminals [R] to [W] with the limit switch open?
A-15, ____V

Q-16, Terminals [T] to [T], lines 14 and 18, thermostat contacts open?
A-16, ____V

Q-17, Burner relay coil, line 18, with the heating thermostat contacts closed? (Flame)
A-17, ____V

Q-18, Safety heater, line 17, with the heating thermostat contacts closed and no flame?
A-18, ____V

Q-19, Terminals [FD] to [FD], lines 15 and 17, with the heating thermostat open and the cad cell removed?
A-19, ____V

Q-20, Terminals [C] to [R], lines 4 and 5, on the fan motor with the plenum switch open?
A-20, ____V

List, for each problem, all of the electrical items that could be checked for defect.

Q-1, The burner is cycling on the limit switch.
A-1, ________________________________

Q-2, The pump motor is running, but there is no 10,000-volt spark from the ignition transformer.
A-2, ________________________________

Q-3, The pump motor is not operating, the burner contacts are closed, there is a 10,000 volt spark at the electrodes, and there is no flame.
A-3, ________________________________

Q-4, There is no voltage from terminals [R] to [W].
A-4, ________________________________

Q-5, You want to check the safety heater circuit.
A-5, ________________________________

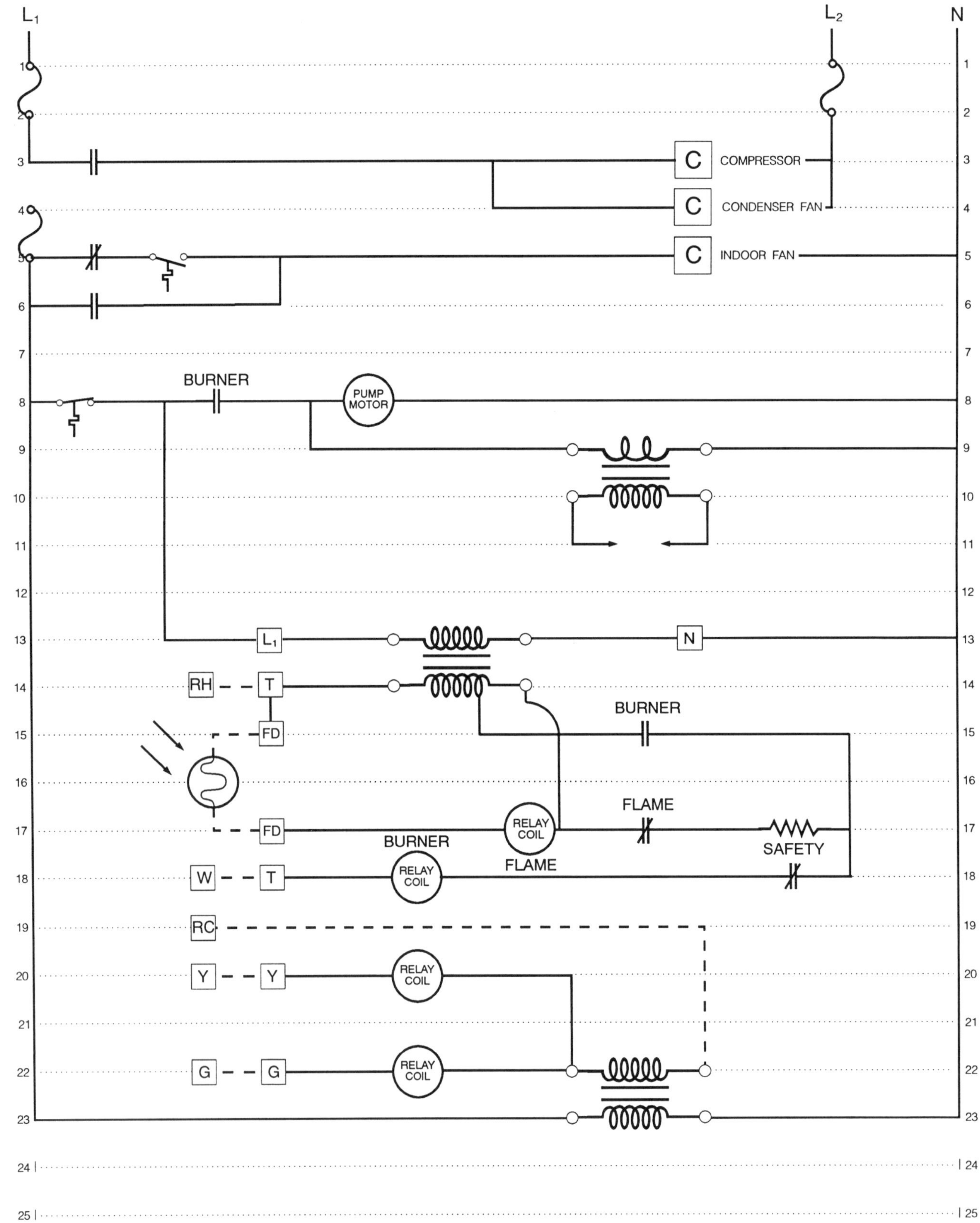
L1
L2
N
C COMPRESSOR
C CONDENSER FAN
C INDOOR FAN
BURNER
PUMP MOTOR
L1
N
RH
T
FD
BURNER
FD
RELAY COIL
FLAME
SAFETY
BURNER
W
T
RELAY COIL
FLAME
RC
Y
Y
RELAY COIL
G
G
RELAY COIL

Example 5
Oil-fired, Forced-Air Furnace with Automatic Air Conditioning

This example shows how automatic air conditioning is added onto the oil-fired, forced-air furnace shown in Example 4. Since the furnace schematic has been fully developed, this example will begin with the terminal center that links these two complementary systems.

1. Define the Function of the Equipment

This is an automatically-operated air conditioning system that works in conjunction with an existing oil-fired, forced-air system.

2. List the Electrical Components

- Fuses are required by the Safety Code.
- A thermostat, with terminals [RH] and [RC] (common), [W] (heat), [Y] (cooling) and [G] (fan), is required for two transformer automatic control.
- A fan relay coil (RELAY COIL) and one set of normally-closed —|/|— and one set of normally- open —||— contacts is needed for fan operation during either heating or cooling.
- An internally-protected, permanent split capacitor compressor motor and condenser fan motor are both needed for cooling.
- A contactor relay coil (RELAY COIL) and a set of contacts —||— is needed to start the compressor and condenser fan motors.
- A step-down transformer is needed to supply 24-V power to the fan and cooling circuits.

3. Go Step-by-Step Through the Electrical Sequence

- When cooling is desired, the manual mode switch on the thermostat must be set on COOL. The manual fan control switch is usually on AUTO.
- When the thermostat closes on a temperature rise, a circuit is completed between terminals [RC] and [Y]. This completes a circuit through the 24-V side of the step-down transformer to the contactor coil (RELAY COIL). The contactor coil closes the contacts —||— and completes parallel circuits to the common connection of the compressor and the condenser fan motors.
- At the same time, another circuit is completed from [RC] to [G], through the indoor fan relay coil (RELAY COIL). When the coil is energized, it opens the normally-closed set of contacts —|/|— and closes the normally-open set of contacts —||—. As a result, the normally-open fan switch is bypassed and a circuit is completed to the common connection of the indoor fan motor.

4. DRAW THE SCHEMATIC

STEP 1. Redraw the burner primary control from Example 4, placing L_1 on line 13, T on lines 14 and 18, FD on lines 15 and 17. Fill in the remaining elements.

STEP 2. Draw the thermostat terminals: RH on line 14, W on line 18, RC on line 19, Y on line 20 and G on line 22.

STEP 3. Field wire - - - the burner primary control to the thermostat by connecting RH to T, W to T.

On line 16, draw the flame detector and field wire it to the FD terminals.

STEP 4. Connect the line voltage side of a step-down transformer between L_1 and N on line 23. Connect the 24-V side of the transformer to an indoor fan relay coil on line 22 and connect the coil to terminal G. Draw a contactor coil on line 20. Connect it to terminal Y and to the 24-V transformer on line 22. Connect the return side of the transformer to terminal RC on line 19. Connect Y and G to Y and G on the thermostat.

STEP 5. On line 5, draw a normally-closed set of indoor fan relay contacts and connect them to L_1 and a plenum switch Connect the plenum switch to terminal C of a PSC indoor fan motor. (Due to the lack of space, motors in this example will be indicated by a terminal C, even though the internal protector, the start and run windings, and the run capacitor have been previously shown.) On line 6, draw a normally-open set of indoor fan relay contacts and connect them to L_1 and to a point on line 5 between the plenum switch and terminal C.

STEP 6. Insert a fuse between lines 1 and 2, to protect L_2. The outdoor compressor and condenser fan motor are 240V.

STEP 7. On line 3, draw a set of compressor contactorss between L_1 and terminal C of a PSC compressor motor.

STEP 8. On line 4, draw terminal C of a PSC condenser fan motor and connect it to line 3 between the contactor and the compressor.

STEP 9. Starting on line 8, redraw the oil burner high side from Example 4. Indicate the junction of the burner primary transformer.

The tear-out
schematic layout sheet
can be found on page 233.

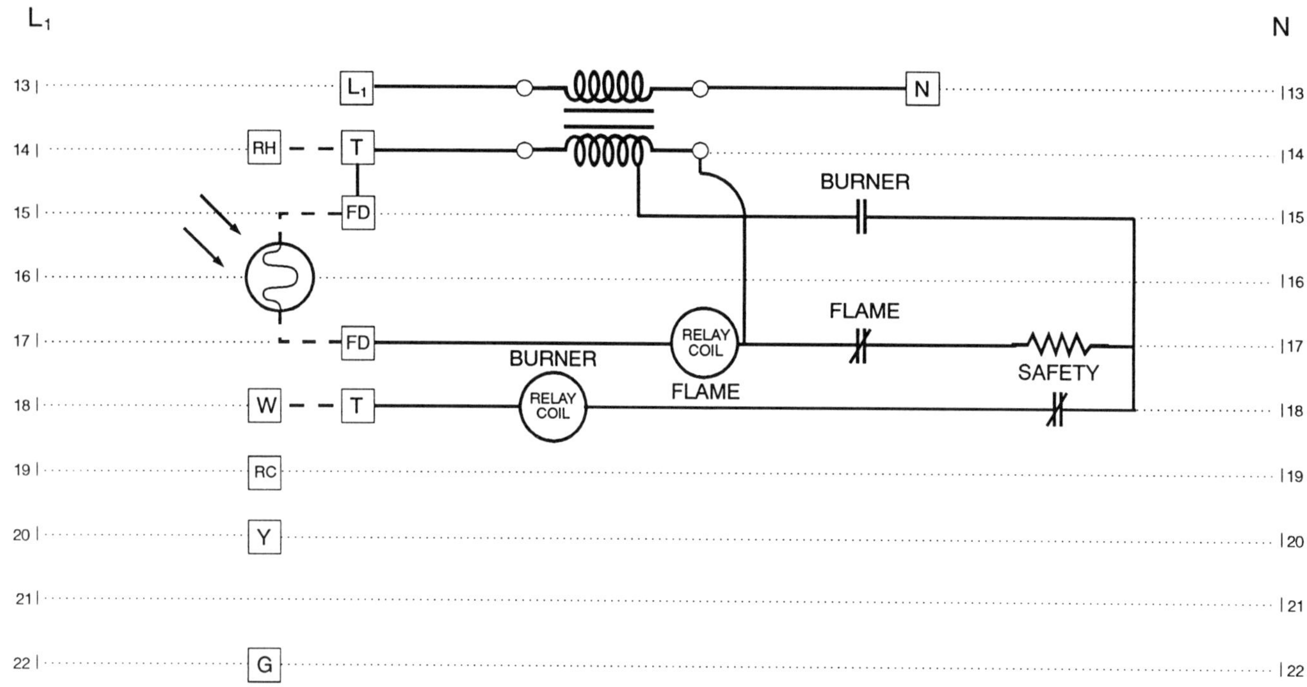

STEPS 1, 2 AND 3

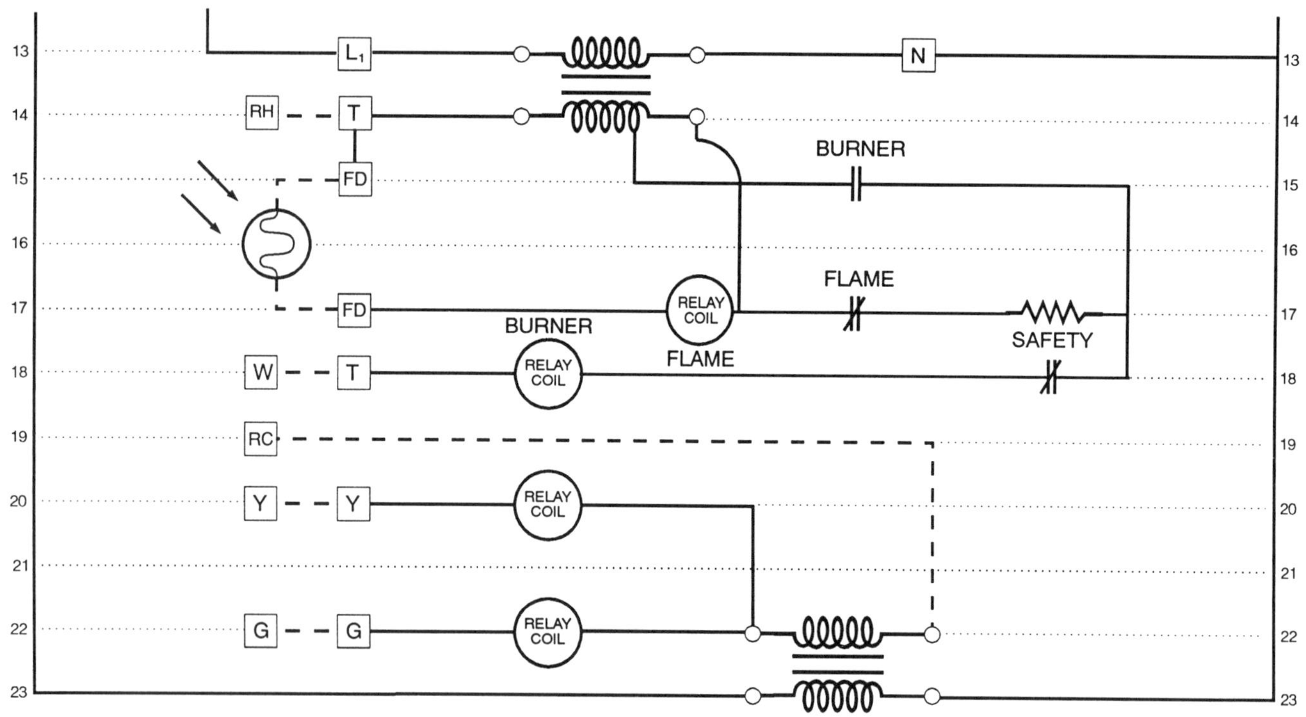

STEP 4

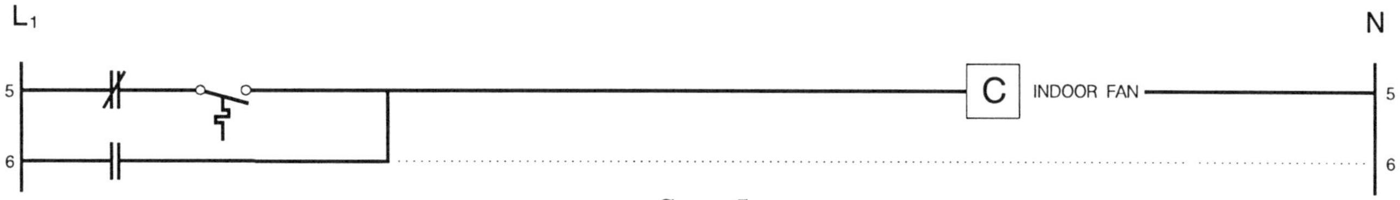

STEP 5

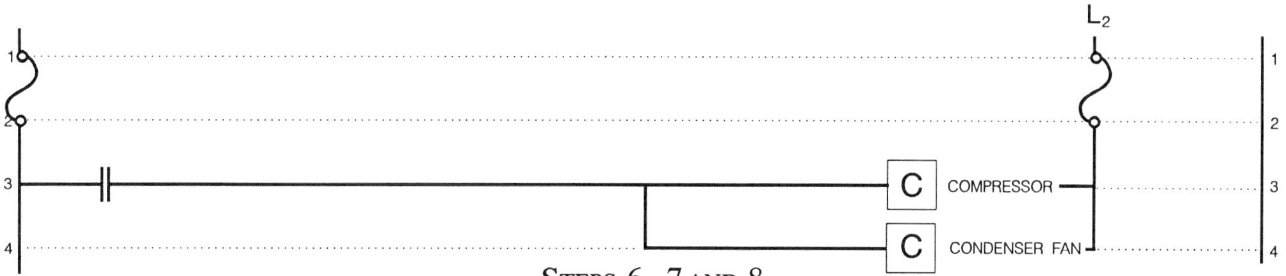

STEPS 6, 7 AND 8

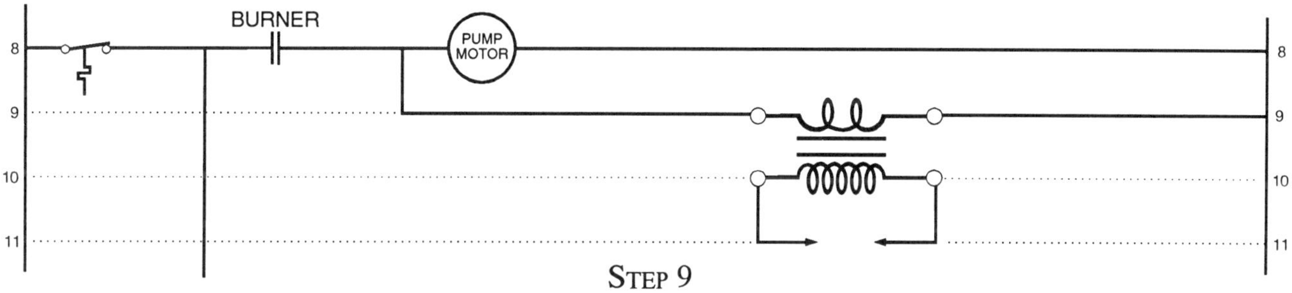

STEP 9

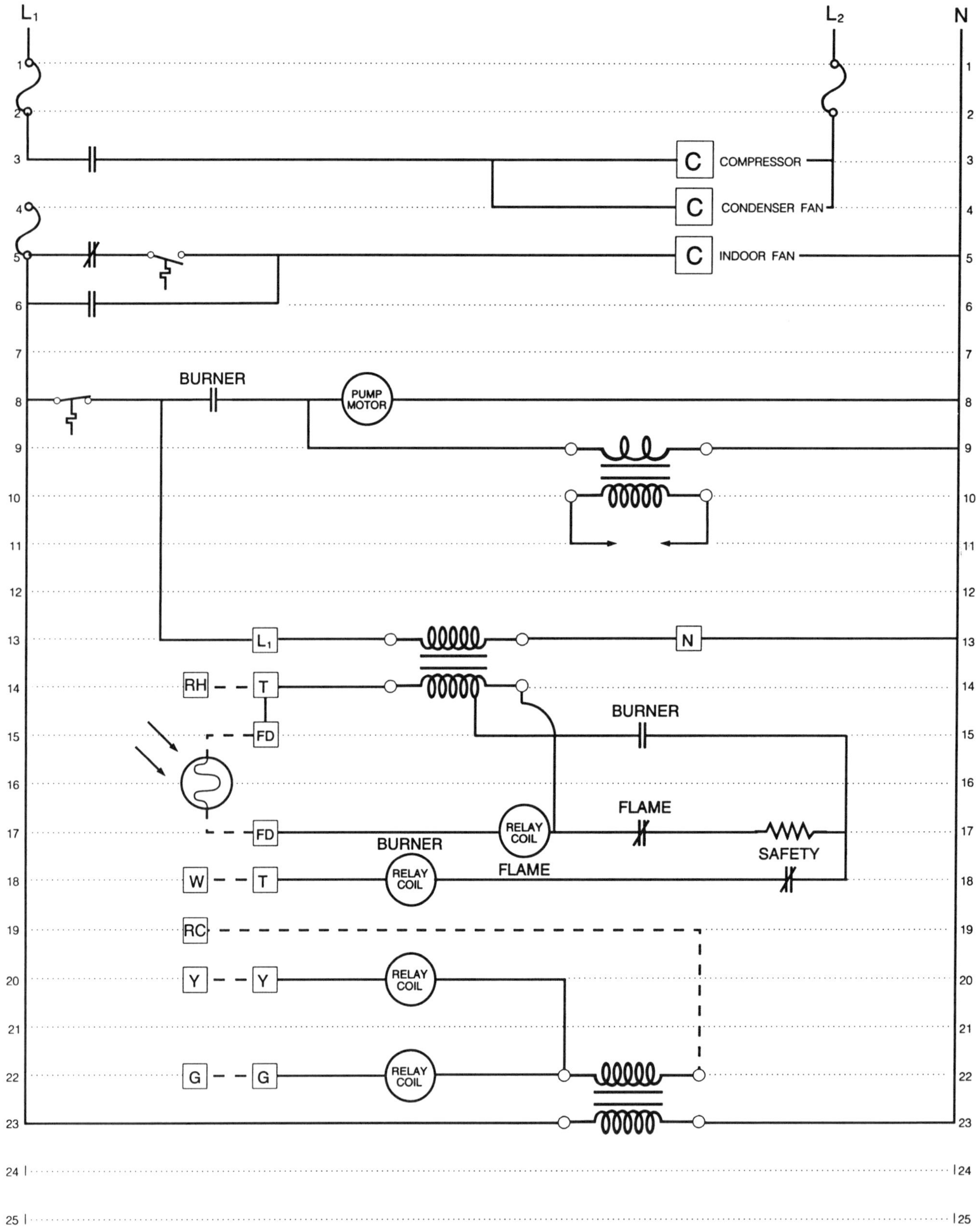
L1
L2
N
C
COMPRESSOR
C
CONDENSER FAN
C
INDOOR FAN
BURNER
PUMP MOTOR
L1
N
RH
T
FD
BURNER
FD
RELAY COIL
FLAME
BURNER
SAFETY
W
T
RELAY COIL
FLAME
RC
Y
Y
RELAY COIL
G
G
RELAY COIL

QUESTIONS FOR EXAMPLE 5

What voltages would you expect to find at the following contacts and loads. Read the question carefully before answering.

Q-1, Condenser contactor coil, with cooling system operating?

A-1, ____V

Q-2, Fan relay coil, line 22, with the cooling system operating?

A-2, ____V

Q-3, Limit switch, with the cooling system operating?

A-3, ____V

Q-4, Terminals [R] to [Y], lines 19 and 20, with the system thermostat on heat and the heating contacts closed?

A-4, ____V

Q-5, Indoor fan running, thermostat in the cooling position with the cooling contacts closed, what do you read across the normally open fan relay contacts, line 6?

A-5, ____V

Q-6, With the indoor fan running and the thermostat contacts [R] to [Y] closed, what do you read across the normally closed fan relay contacts, line 5 ?

A-6, ____V

Q-7, Terminals [T] to [T] with the cooling system operating?

A-7, ____V

Q-8, Terminals [R] to [W] with the thermostat selector on heat and the heating system operating? (Flame)

A-8, ____V

Q-9, Cooling low-voltage transformer primary with the system selector switch in the OFF position?

A-9, ____V

Q-10, Heating transformer secondary with the thermostat selector switch in the OFF position?

A-10, ____V

Answer the following questions. List, for each problem, all of the electrical items that could be checkcd for defect.

Q-1, The compressor is operating but the indoor evaporator fan motor is off.

A-1, ______________________________

Q-2, The evaporator fan motor is running but the compressor and condenser fan motor will not come on.

A-2, ______________________________

Q-3, The indoor fan runs when the system is heating, but does not operate with the system on cooling.

A-3, ______________________________

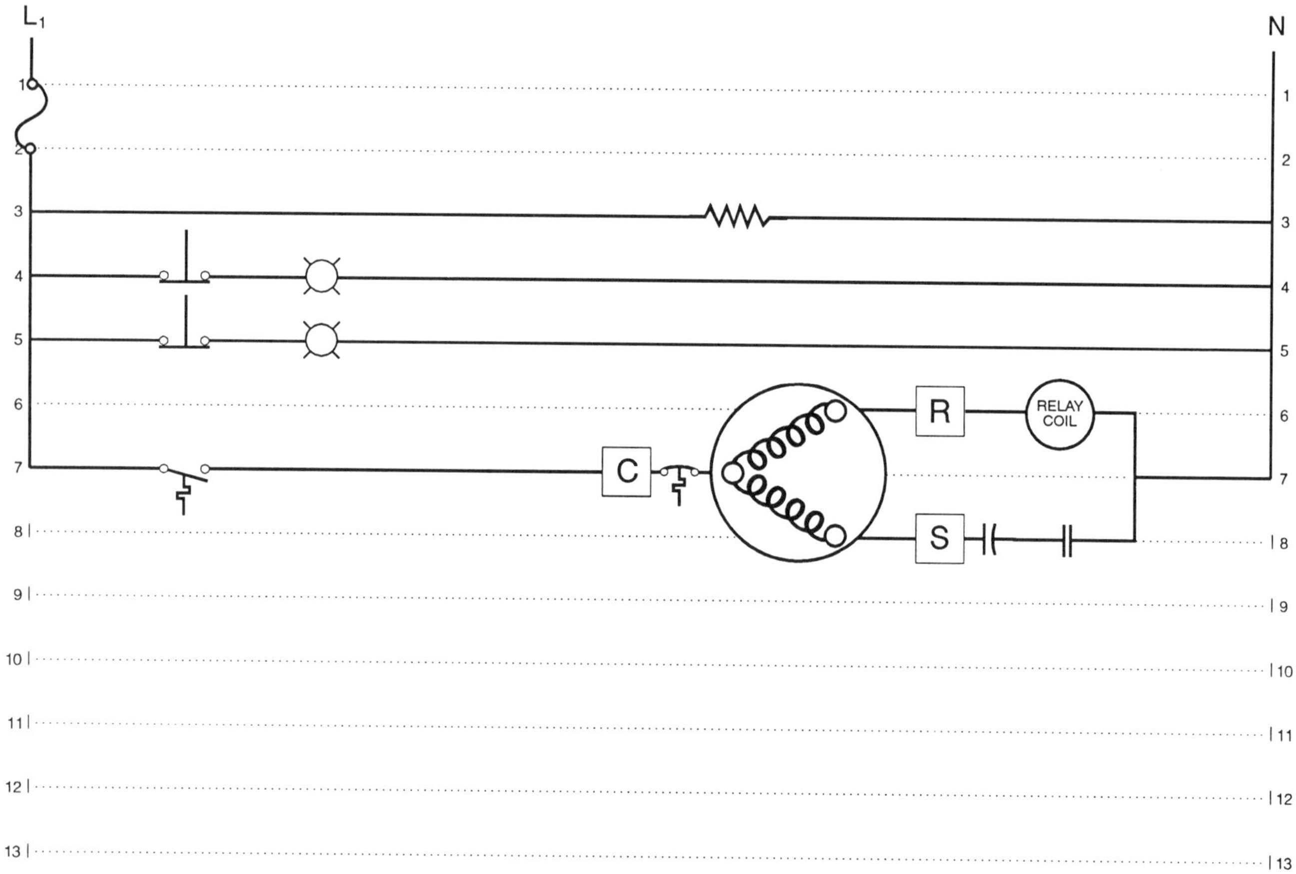
L1
N
1
2
3
4
5
6
7
8
9
10
11
12
13
R
RELAY COIL
C
S

EXAMPLE 6 TWO-DOOR REFRIGERATOR / FREEZER

The domestic refrigerator/freezer is delivered as a complete unit and only needs to be plugged in to begin operating. This example will show only the basic electrical circuits used in low-cost units that do not feature automatic defrost.

1. DEFINE THE FUNCTION OF THE EQUIPMENT

This is an automatically-operated, refrigerator/ freezer that runs on line voltage and is equipped with a freezer door heater to prevent condensate from freezing the door shut.

2. LIST THE ELECTRICAL COMPONENTS

The standard electrical source is 120 V, 60 cycle (60 Hz), single-phase (1φ), alternating current (ac).

- A fuse is required by the Safety Code.
- A freezer door heater is needed to prevent freezing of condensate at the door.
- Light bulbs are needed in both the refrigerator and freezer sections for greater visibility. They will be controlled by normally-closed push button switches that open when the door is closed.
- A compressor is needed for cooling. It will have a high torque motor, equipped with a current (magnetic) relay.
- A cooling thermostat is needed for temperature control.

3. GO STEP-BY-STEP THROUGH THE ELECTRICAL SEQUENCE

- Line voltage (120 vac) must be supplied to all components.
- There is no switch controlling the freezer door heater, therefore is produces heat as long as power is available.
- Lights in the freezer and refrigerator compartments are controlled by normally-closed push button switches that close when the door is open, completing a circuit to the light bulbs.
- Cooling is provided by a refrigerant compressor whose internally-protected motor is equipped with a current relay for extra starting torque. When the motor attains approximately 80% of full speed, the starting relay drops out of the circuit.
- The compressor is cycled on and off by a cooling thermostat to maintain a preset temperature.

4. DRAW THE SCHEMATIC

STEP 1. Draw in the line power (120 V) using the two vertical lines L_1 and N. Insert a fuse in L_1 between lines 1 and 2.

STEP 2. On line 3, draw the freezer door heater.

STEP 3. On lines 4 and 5, draw a pushbutton switch and connect it to a light bulb.

STEP 4. On line 7, draw a cooling thermostat and connect it to an internally-protected split-phase motor. Draw a current relay, consisting of a coil on line 6 and a pair of normally-open contacts on line 8, between the run and start windings and connect it to N.

STEP 5. Connect all of the elements thus drawn to L_1 and N.

The tear-out schematic layout sheet can be found on page 235.

STEP 1

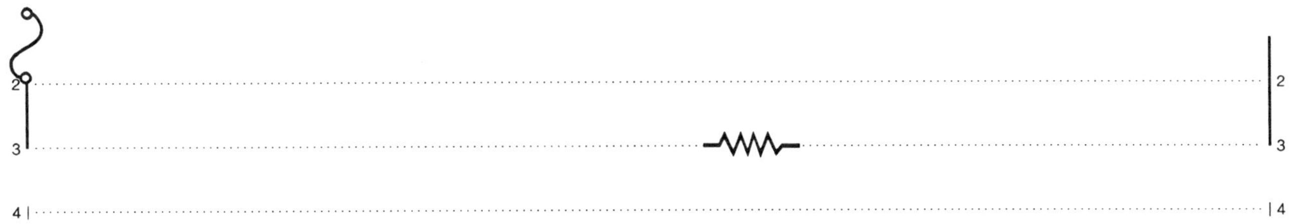

STEP 2

STEP 3

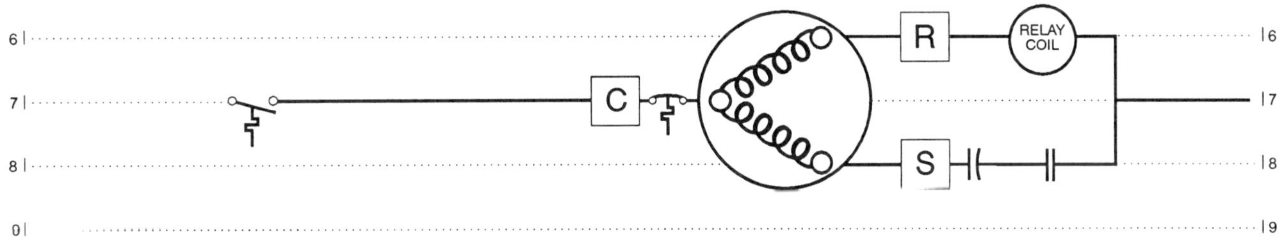

STEP 4

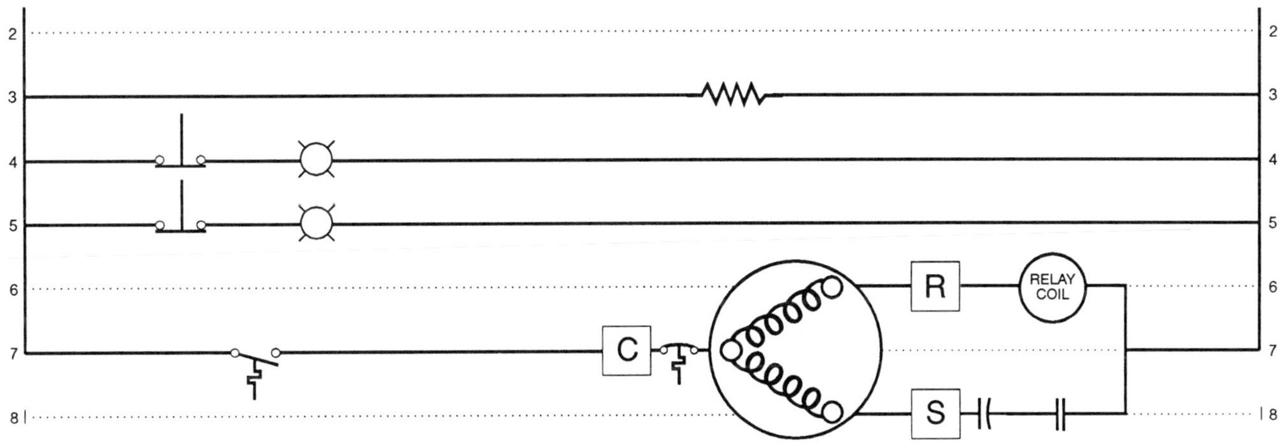

STEP 5

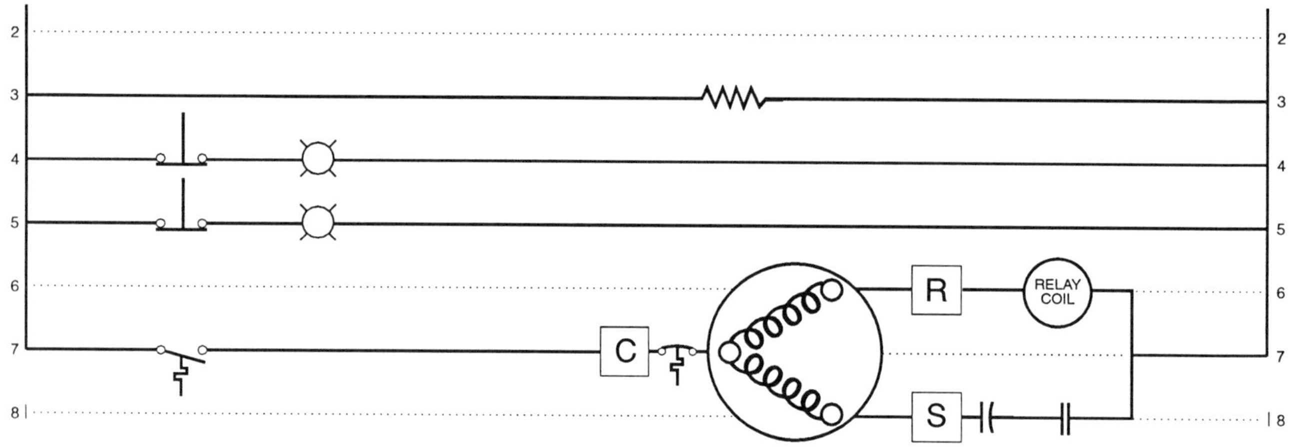

QUESTIONS FOR EXAMPLE 6

What voltages would you expect to find at the following contacts and loads? Read the questions carefully before answering.

Q-1, The freezer door heater, line 3 ?
A-1, ____V

Q-2, The freezer light with the door open?
A-2, ____V

Q-3, The refrigerator light with the door open?
A-3, ____V

Q-4, The freezer light door switch with the door open?
A-4, ____V

Q-5, The refrigerator light door switch with the door open?
A-5, ____V

Q-6, The thermostat contacts with the compressor running?
A-6, ____V

Q-7, The thermostat contacts with the compressor off?
A-7, ____V

Q-8, Compressor terminals [R] to [C] with the compressor running?
A-8, ____V

Q-9, What would you expect to read across the freezer door switch with the door closed?
A-9, ____V

Q-10, What would you expect to read across the refrigerator light with the door closed?
A-10, ____V

Answer the following questions. List, for each problem, all of the electrical items that could be checked for defect.

Q-1, The freezer light does not light with the door open.

A-1, ______________________________

Q-2, The compressor does not start or run with the thermostat contacts closed.

A-2, ______________________________

Q-3, The freezer door is frozen shut.

A-3, ______________________________

Q-4, Nothing works, no lights, compressor or heater.

A-4, ______________________________

Q-5, There is no voltage to the compressor terminals R and C, but the lights and door heater operate.

A-5, ______________________________

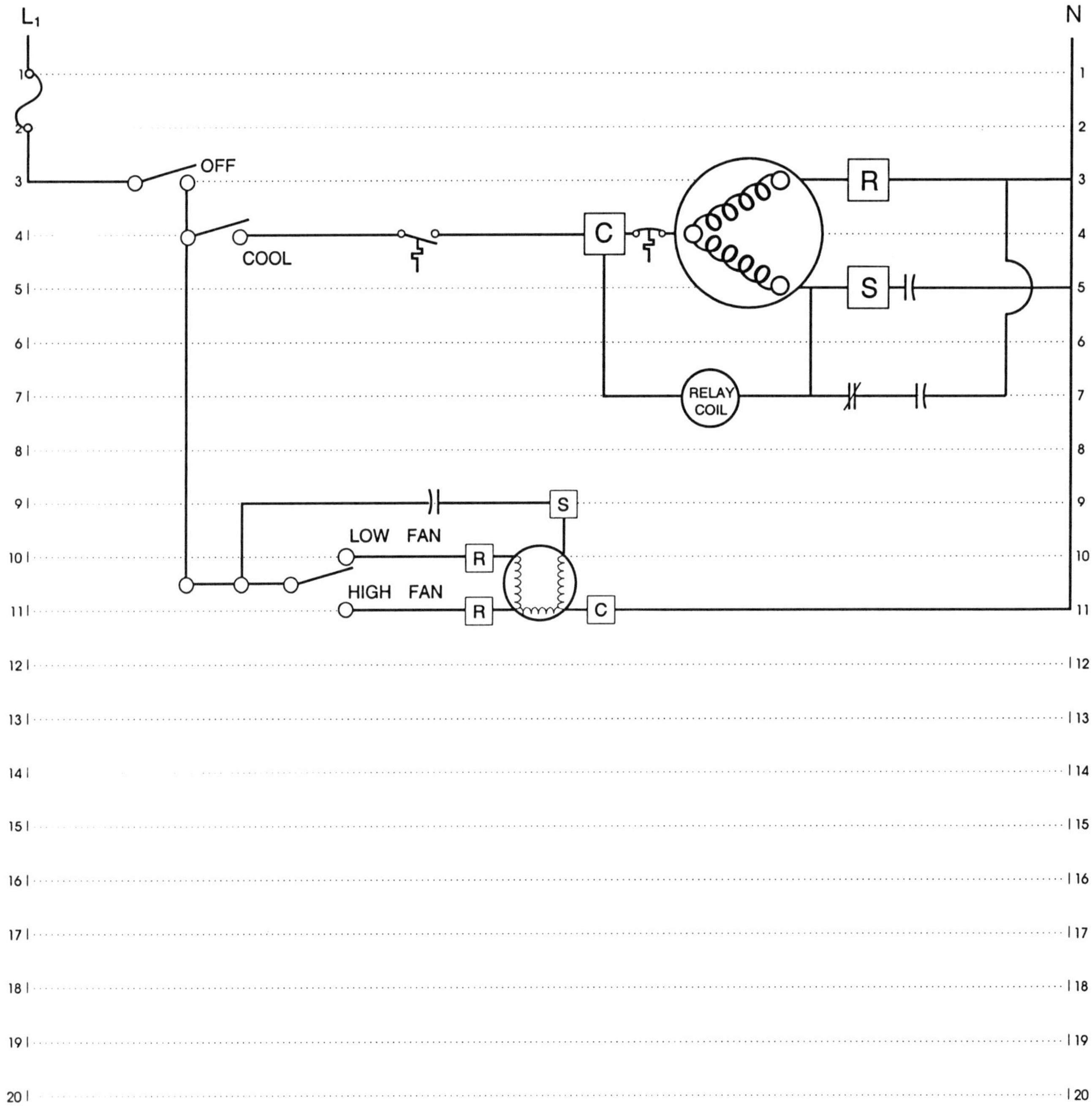

L1
N
OFF
COOL
C
R
S
RELAY COIL
LOW FAN
HIGH FAN
R
S
R
C

EXAMPLE 7
WINDOW AIR CONDITIONER

The typical window air conditioner is a plug-in unit that is completely wired by the manufacturer. While some window air conditioners have very involved electrical circuits, this example will cover only the basic, two-speed model.

One feature of note is the fan motor. This motor has a double-end shaft that drives two fans: one to circulate the indoor air, the other to cool the refrigerant condenser.

1. DEFINE THE FUNCTION OF THE EQUIPMENT

This is an automatically-operated air conditioner with a two-speed motor that can be operated even when cooling is not required.

2. LIST THE ELECTRICAL COMPONENTS

- The standard home electrical source is 120 V, 60 Hz, (1ϕ), alternating current (ac).
- A fuse  is required by the Safety Code.
- A selector switch is needed for mode control. This multi-position switch selects the following operating modes:

SWITCH POSITION	CONTACTS MADE
OFF	None
HI FAN	L_1-HI
LO FAN	L_1-LO
HI-COOL	L_1-HI-COOL
LO-COOL	L_1-LO-COOL

- A two-speed fan motor 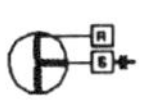 is required for LO and HI FAN and COOL.
- A thermostat  is required for automatic control of the cooling compressor.
- A permanent split capacitor motor, with internal protector 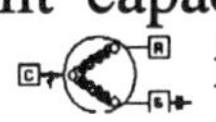 is needed to drive the compressor.
- In areas that experience persistent low voltage, a hard-start kit may be required.

3. GO STEP-BY-STEP THROUGH THE ELECTRICAL SEQUENCE

- Line voltage (120 vac) must be supplied to the unit.
- Cycling of the cooling compressor is controlled by a cooling thermostat when the manually-operated selector switch is on COOL.
- The compressor motor is an internally-protected PSC type with an optional hard-start kit composed of a start capacitor and a potential relay.
- The two-speed fan motor is essentially a PSC type with a run capacitor and an added low speed coil. It is controlled by a manually- operated mode switch that allows it to operate independently of the cooling compressor.

4. DRAW THE SCHEMATIC

STEP 1. Insert a fuse in L_1, between lines 1 and 2.

STEP 2. Draw single-pole, single-throw switches on lines 3 and 4 and connect them in series.

STEP 3. Connect the switch on line 4 to a cooling thermostat and terminal C of a PSC motor. Connect terminals R and S to N.

STEP 4. Draw the potential relay coil on line 7. Connect the relay coil to terminal C and to a normally-closed switch and a capacitor. Connect the capacitor to the RUN circuit and the relay coil to the START circuit.

STEP 5. Draw a two-speed motor with the R terminals on lines 10 and 11, the S terminal on line 9. Connect terminal C to N.

STEP 6. Draw a single-pole, double-throw switch with its terminals on lines 10 and 11. Connect its common terminal to the OFF switch on line 3. Connect the switch terminals to the two-speed motor and label the lines HI FAN and LO FAN.

The tear-out
schematic layout sheet
can be found on page 237.

STEP 1

STEP 2

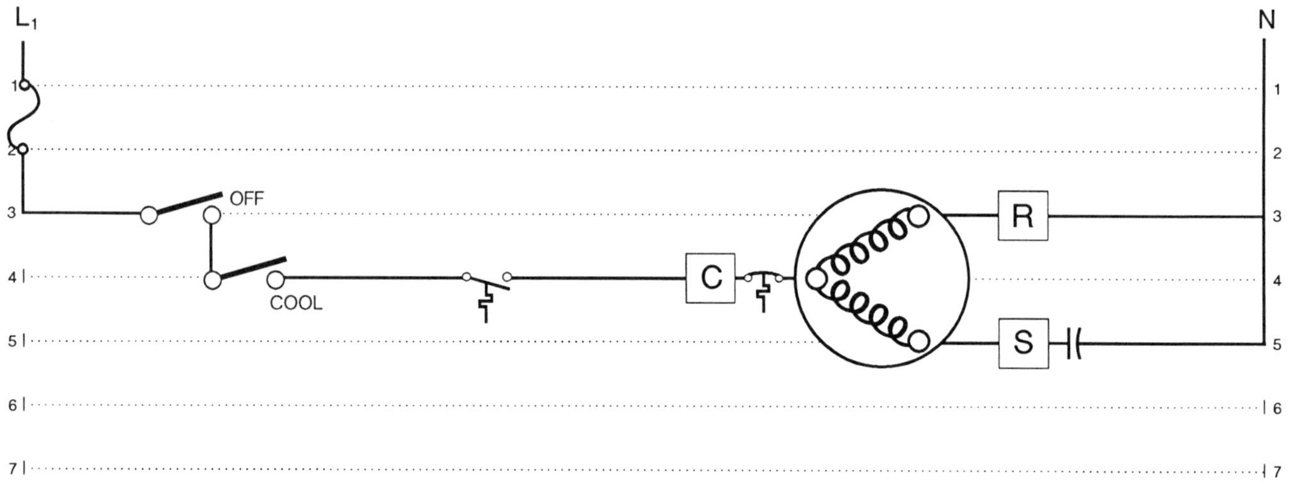

STEP 3

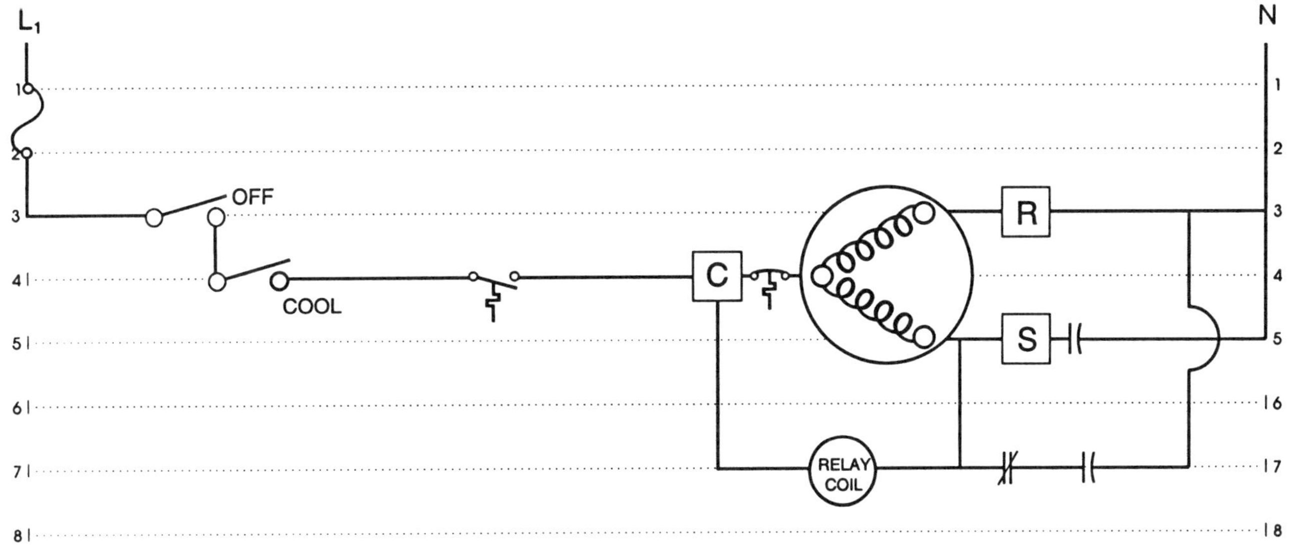

STEP 4

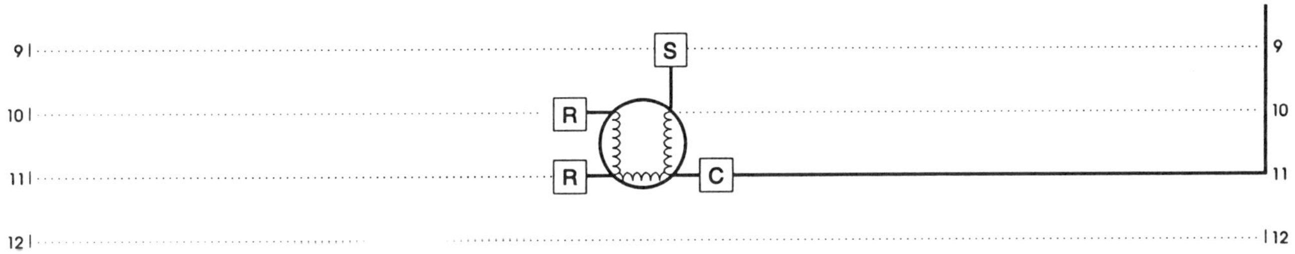

STEP 5

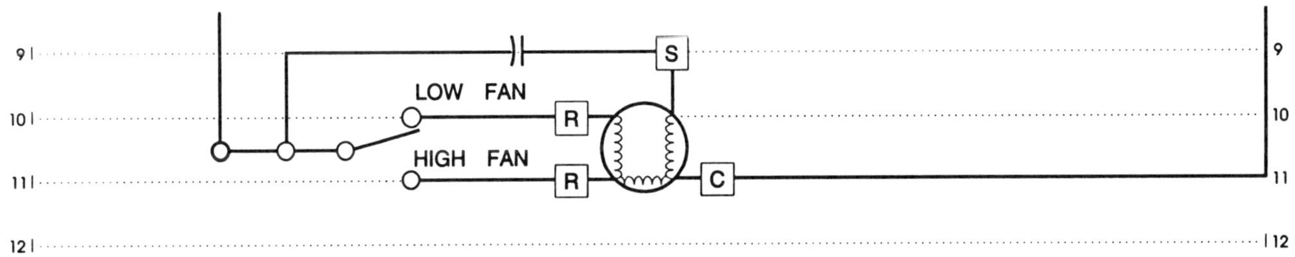

STEP 6

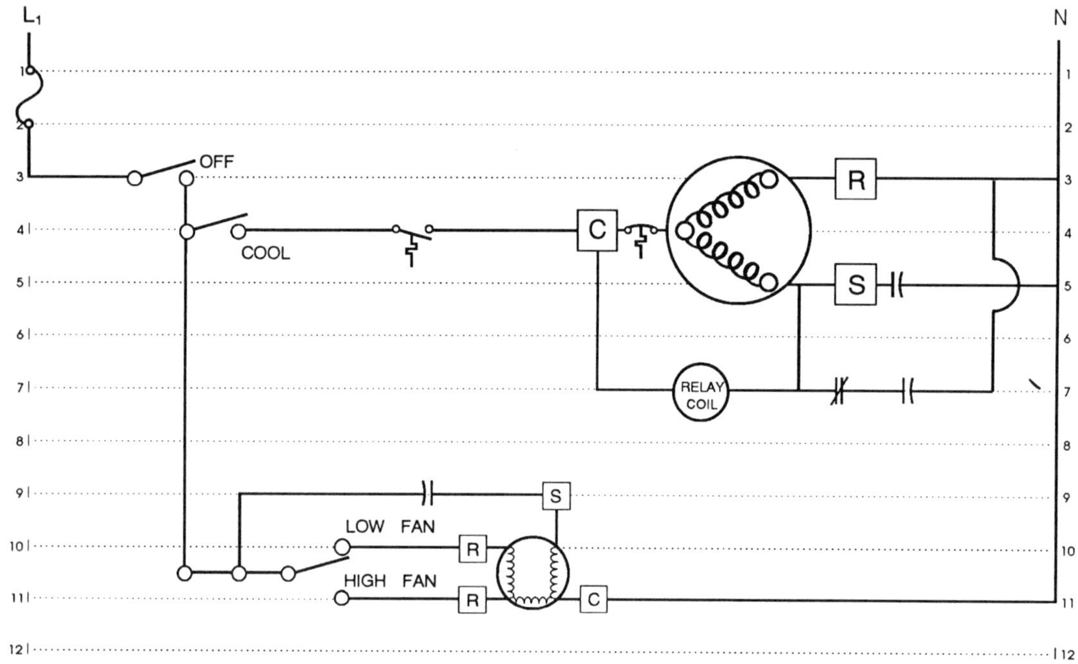

QUESTIONS FOR EXAMPLE 7

What voltages would you expect to find at the following contacts and loads? Read the questions carefully before answering.

Q-1, Cool switch contacts, line 4, with the system switch off?
A-1, ____V

Q-2, Thermostat contacts, line 4, with the switch off?
A-2, ____V

Q-3, Thermostat contacts with the compressor running?
A-3, ____V

Q-4, Compressor terminals C to R with the compressor running?
A-4, ____V

Q-5, Fan motor terminals R to C with the fan running on high speed?
A-5, ____V

Q-6, Fan motor terminals R to C with the fan running on low speed?
A-6, ____V

Q-7, Compressor terminals R to S with the compressor running?
A-7, ____V

Q-8, Fan motor terminals R to S with the fan motor running?
A-8, ____V

Q-9, Off selector switch with the switch open?
A-9, ____V

Q-10, Thermostat with the cooling switch closed and the thermostat satisfied?

List, for each problem, all of the electrical items that could be checked for defect.

Q-1, Only the high-speed fan runs with the system on.

A-1, ________________________________

Q-2, Fan motor runs on either speed but the compressor does not start or run.

A-2, ________________________________

Q-3, The fuse is good but neither the compressor nor fan motor will run.

A-3, ________________________________

Q-4, On low cool, only the compressor will run.

A-4, ________________________________

Q-5, On high cool, only the fan motor will run.

A-5, ________________________________

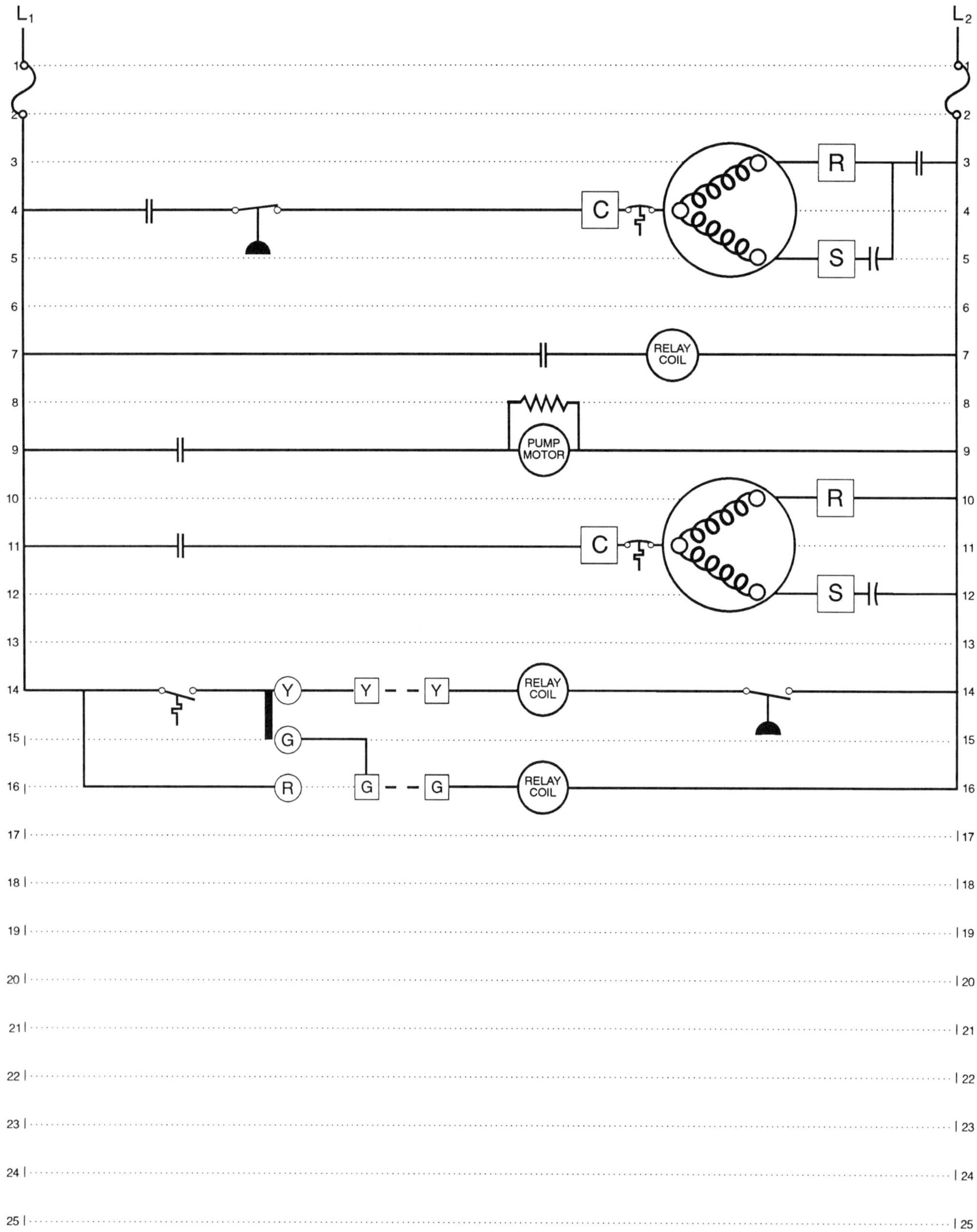

L1
L2
C
R
S
RELAY COIL
PUMP MOTOR
C
R
S
Y
Y
Y
RELAY COIL
G
R
G
G
RELAY COIL

EXAMPLE 8
PACKAGE, AIR CONDITIONER WITH WATER-COOLED CONDENSER

This package unit uses water as a condensing medium. The water pump is mounted in the equipment case and the only piping needed are supply and return water lines to the condenser. The selector switch and thermostat are located on the front of the package and all controls are line voltage.

1. DEFINE THE FUNCTION OF THE EQUIPMENT

This is an automatically-operated air conditioning unit with a water-cooled condenser and line voltage controls.

2. LIST THE ELECTRICAL COMPONENTS

- The primary voltage is 240 V, 60 Hz, 1ϕ, ac.
- Fuses are required by the Safety Code.
- The compressor and indoor fan employ PSC motors. A water pump motor is also required.
- Relays are used to close the contacts for all three motors.
- High-pressure and low-pressure controls are used to protect the compressor.
- Because the water pump must start before the compressor, a time-delay relay with a 60 second delay, is used.
- The selector switch has three positions: OFF (nothing operates); FAN (only the fan runs); and COOL (both the compressor and the fan run). The compressor is cycled by a line- voltage cooling thermostat.

3. GO STEP-BY-STEP THROUGH THE ELECTRICAL SEQUENCE

- Line voltage (240 V) must be supplied to the unit.
- With the selector switch on FAN, the fan relay is energized, closing the contacts to the fan motor.
- With the selector switch on COOL, the fan will run. When the thermostat closes on a call for cooling, the water pump relay coil will be energized, closing the water pump contacts on line 9, if the low pressure control is closed.
- With the water pump running, the time-delay relay heater is energized. Alter 60 seconds it will close the contacts on line 7 and energize the compressor contactor coil, if the high pressure switch is closed. Thus, the compressor cannot start until after the condenser cooling water is circulating.

4. DRAW THE SCHEMATIC

STEP 1. Insert fuses in L_1, and L_2, between lines 1 and 2.

STEP 2. Draw an internally-protected PSC compressor motor with terminal [C] on line 4. Link terminal [C] to a high- pressure switch, a set of contacts —||—, and L_1. On line 3, insert a pair of contacts and join terminal [S] to [R] to L_2.

STEP 3. Insert a set of contacts and wire the water pump between L_1 and L_2 on line 9.

STEP 4. On line 11, draw a set of contacts and wire them between L_1, and terminal [C] of a PSC indoor fan motor. Join terminals [S] and [R] of the motor to L_2.

STEP 5. Draw a thermostat switch on line 14. Show the thermostat connections by field wiring a pair of terminals [Y] on line 14 and [G] on line 16. Draw the internal thermostat contacts: (Y) on line 14, (G) on line 15, and (R) on line 16. Connect (Y) and (G) with a manual slide switch |. Connect (G) to [G] and (R) to L_1 at line 14.

STEP 6. On line 14, connect the water pump relay and a low-pressure switch to [Y] and L_2.

STEP 7. On line 16, connect the indoor fan relay coil to [G] and L_2.

STEP 8. On line 7, insert the time-delay relay contacts —||— and the compressor contactor coil, in series, between L_1, and L_2.

On line 9, draw the pump motor and its contacts —||—.

STEP 9. Add the time-delay relay heater -\/\/\/- on line 8 in a parallel circuit to the water pump and in series with the water pump contactor contacts —||— on line 9.

The tear-out
schematic layout sheet
can be found on page 239.

STEP 1

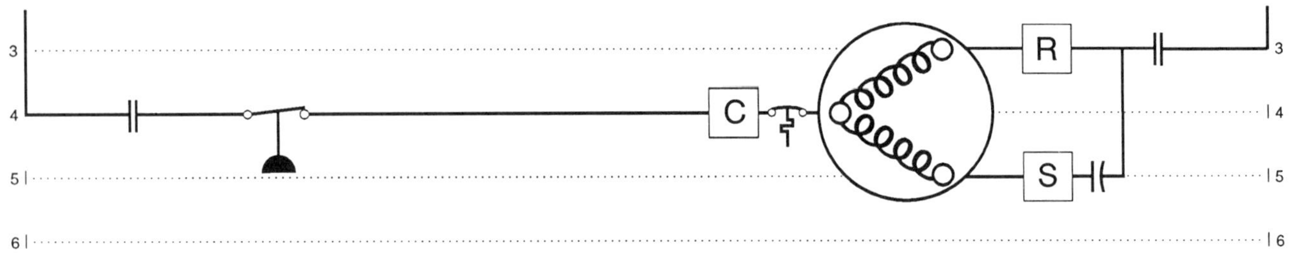

STEP 2

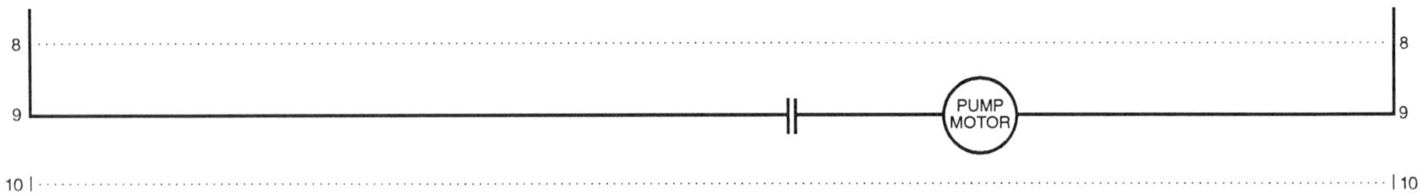

STEP 3

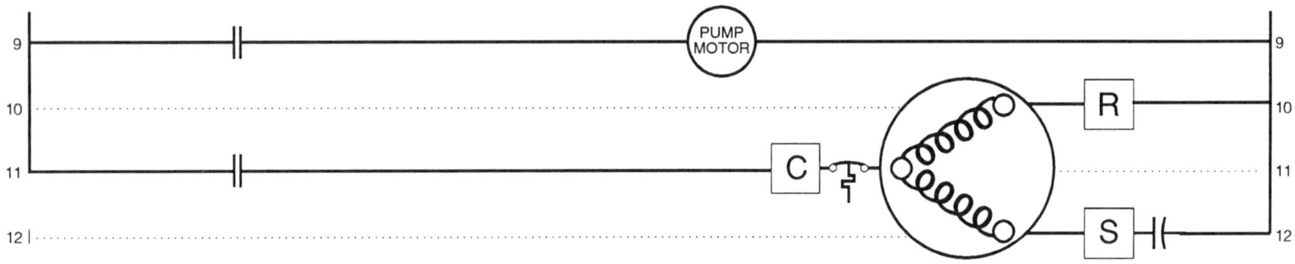

STEP 4

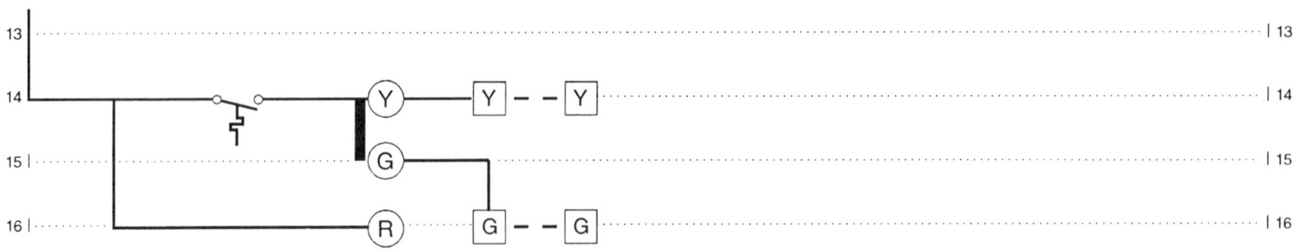

STEP 5

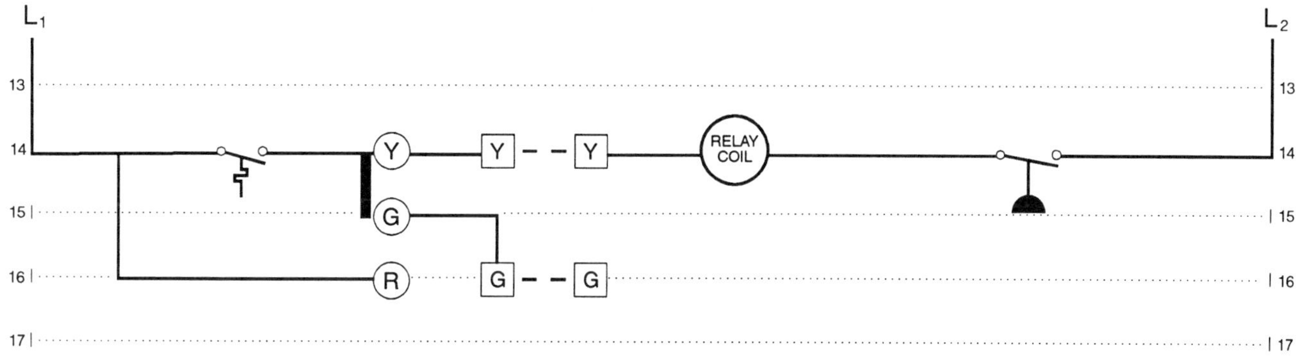

Step 6

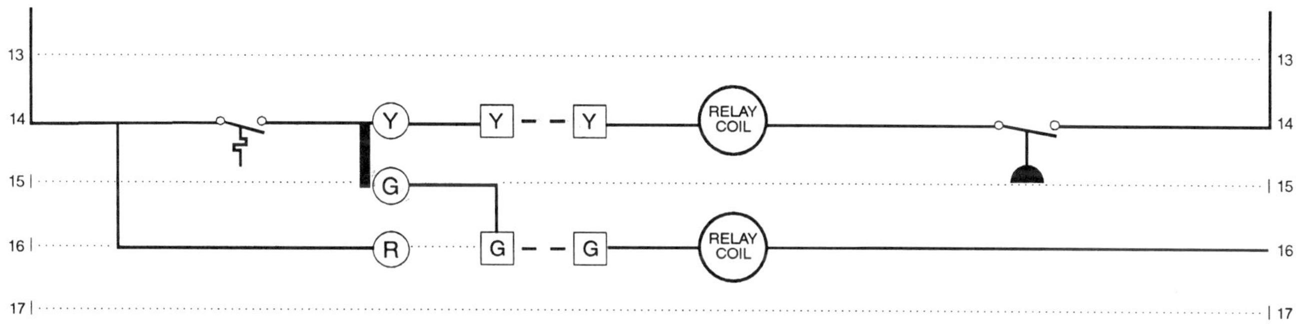

Step 7

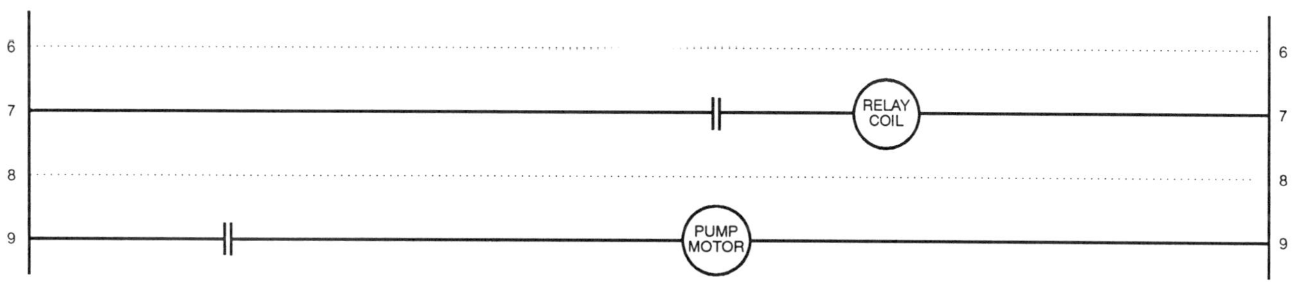

Step 8

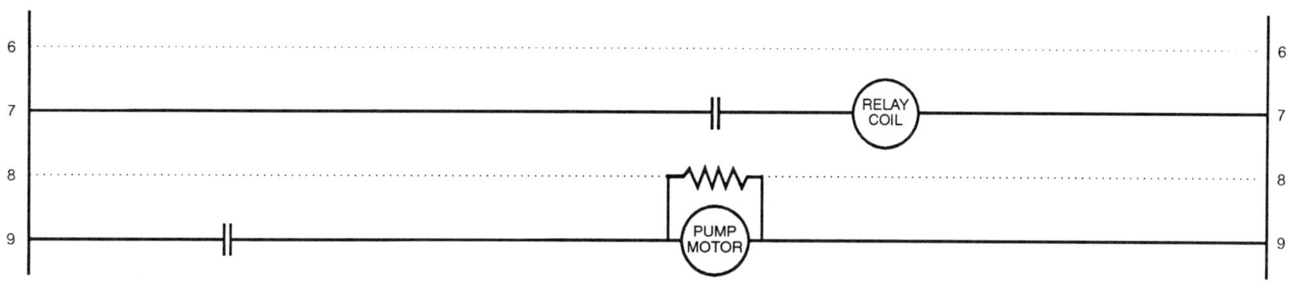

Step 9

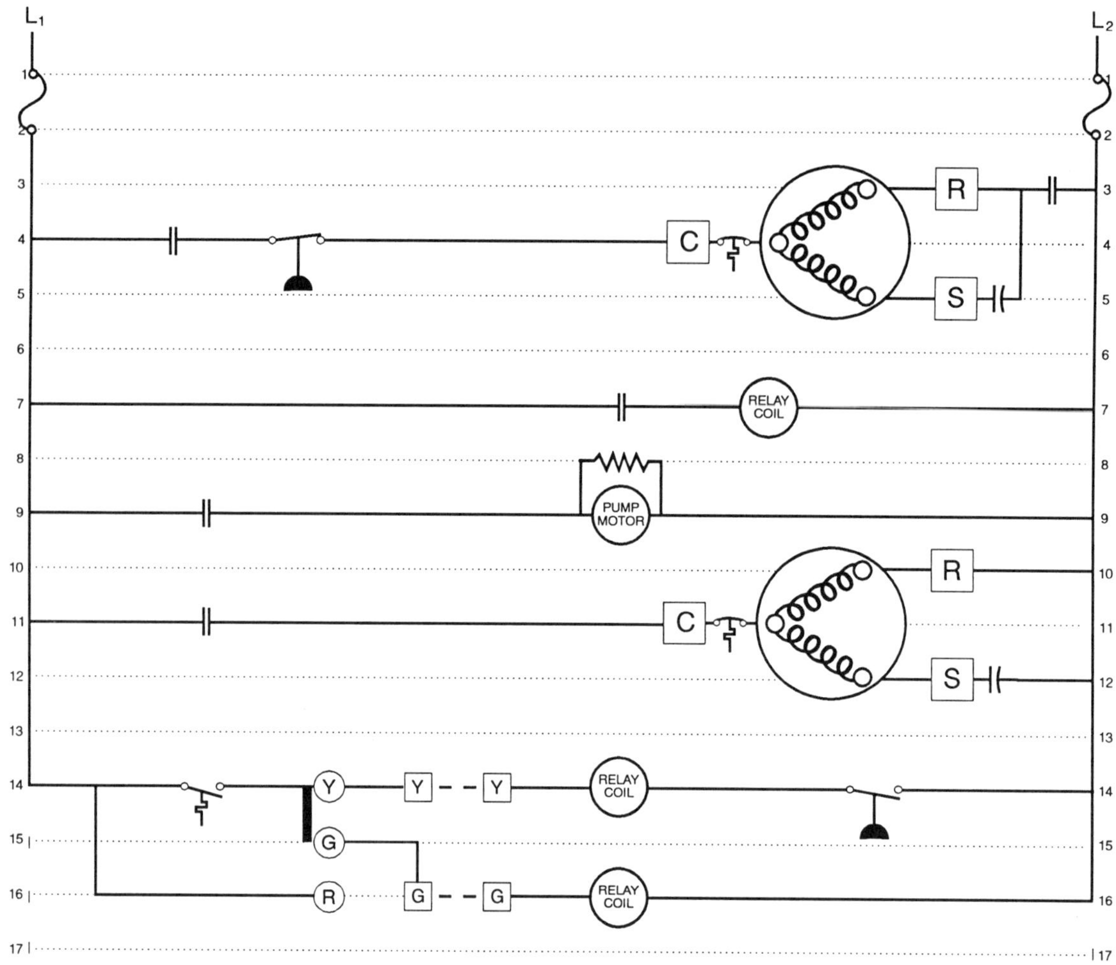

QUESTIONS FOR EXAMPLE 8

What voltages would you expect to find at the following contacts and loads? Read the question carefully before answering.

Q-1, The low pressure switch, line 14, with the system off?

A-1, ____V

Q-2, The low pressure switch with the cooling system operating?

A-2, ____V

Q-3, Each set of contactor contacts, lines 3 and 4, with the compressor running?

A-3, ____V

Q-4, Each set of contactor contacts with the compressor off?

A-4, ____V

Q-5, Relay coil, line 7, with the system in the OFF position?

A-5, ____V

Q-6, Relay coil, line 7, with the compressor running?

A-6, ____V

Q-7, High-pressure switch, line 4, with the system switch in the OFF position?

A-7, ____V

Q-8, Time delay heater, line 8, with the compressor running?

A-8, ____V

Q-9, Pump motor, line 9, with the compressor running?

A-9, ____V

Q-10, Terminals [Y] to [G] with the compressor running?

A-10, ____V

List, for each problem, all of the electrical items that could be checked for defect.

Q-1, Thermostat contacts [R] to [Y] are closed, but the contactor contacts, line 3 and 4, are open.

A-1, ______________________________

Q-2, Compressor is running but the pump motor does not come on.

A-2, ______________________________

Q-3, Compressor is running but the indoor fan motor will not come on.

A-3, ______________________________

Q-4 Pump motor runs, but the compressor will not run.

A-4 ______________________________

Q-5 Compressor will not come on, but you read 240 V from L_1, to L_2, and 0 V from R to C at the compressor terminals.

A-5 ______________________________

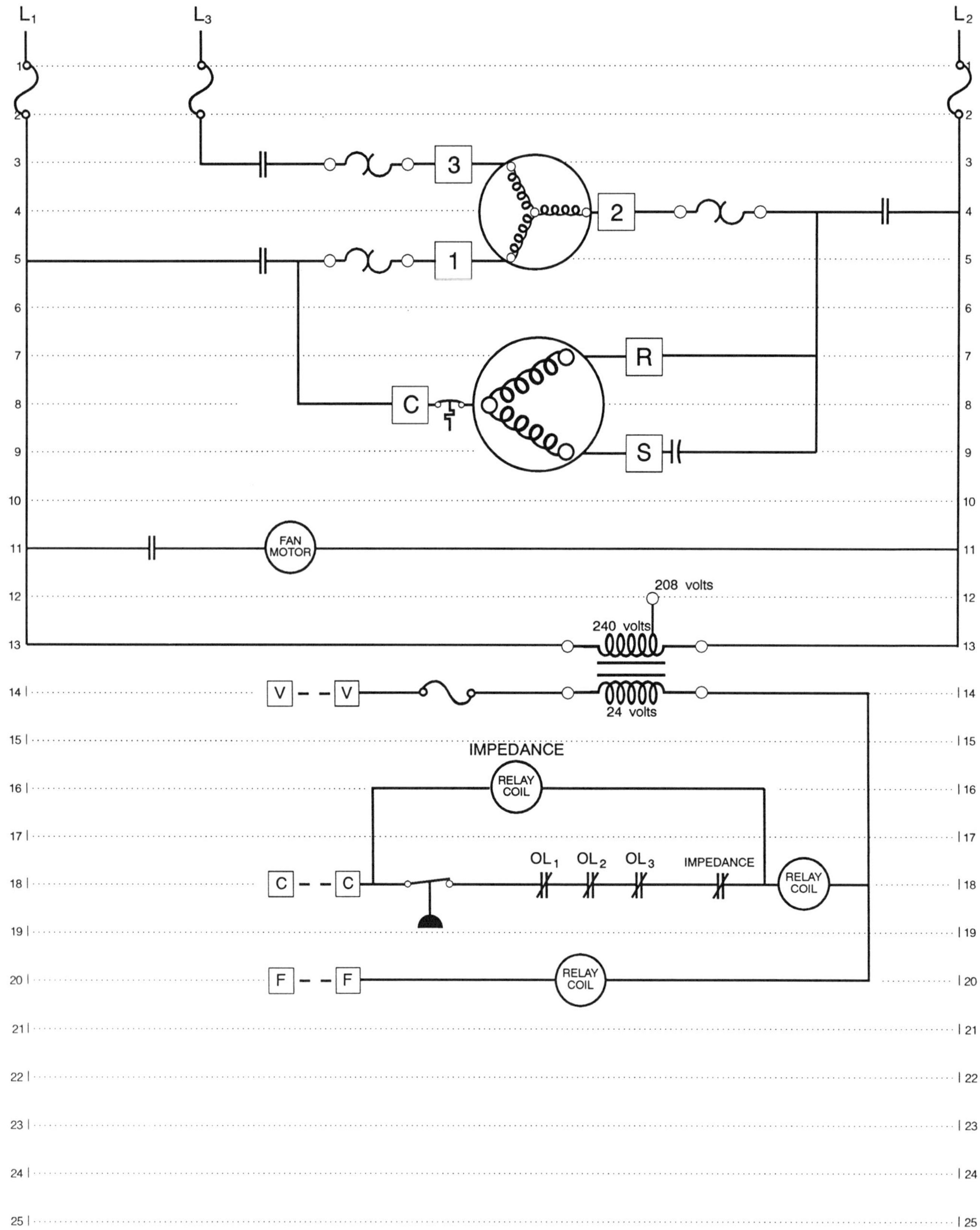
L1
L3
L2
3
2
1
R
C
S
FAN MOTOR
208 volts
240 volts
24 volts
V - - V
IMPEDANCE
RELAY COIL
OL1 OL2 OL3
IMPEDANCE
C - - C
RELAY COIL
F - - F
RELAY COIL

EXAMPLE 9
SINGLE PACKAGE AIR-COOLED AIR CONDITIONER

This package unit is a complete factory-wired system, except for the low-voltage (24 V) remote thermostat. The primary voltage to the unit is 240 V/208 V, 3ϕ, 60 Hz.

1. DEFINE THE FUNCTION OF THE EQUIPMENT

This is an automatically-controlled air conditioner with a 3ϕ compressor.

2. LIST THE ELECTRICAL COMPONENTS

- The electrical source is 240 V/208 V, 60 Hz, 3ϕ, ac.
- Fuses are required by the Safety Code.
- A 3ϕ motor with thermal protectors is required for cooling.
- A contactor, consisting of a relay coil and contacts in each of the three-phase lines, is needed to start the compressor.
- Each of the three-phase leads should be protected by an overload OL_1 in the 24-V cooling circuit.
- Additional protection is provided by a high-pressure switch.
- An impedance relay is to be used to prevent automatic restarting after a safety device has opened.
- A single-phase, PSC motor is needed to cool the condenser. It is operated in tandem with the compressor.
- An indoor fan motor is needed to circulate the cooled air. A fan relay controls this motor.
- A 240-V step-down transformer, with a 208-V tap, is needed to supply 24-V current to the low-voltage circuit. The low side of the transformer is protected by a fuse.
- A thermostat, with terminals V, C, and F is needed for automatic control.

3. GO STEP-BY-STEP THROUGH THE ELECTRICAL SEQUENCE

- Set the thermostat on COOL, fan on AUTO, and lower the temperature setting to close contacts V to C and F.
- With all of the safety control contacts, OL_1 closed, the contactor coil and the indoor fan relay coil will be energized. Three sets of contacts will close, starting the compressor and the outdoor fan motor.
- The indoor fan motor will also start when its contacts close.
- If any safety control opens, only the compressor and the outdoor fan motor will stop.
- The opening of the thermostat contacts V to F will stop the indoor fan motor. If any safety device opens, the cause of the interruption must be determined and corrected, and the impedance relay manually reset, before normal operation can resume.

4. DRAW THE SCHEMATIC

STEP 1. Start with the high-voltage (240 V) section. Draw in lines L_1, L_2, L_3 above line 1. Fuse each of the power lines between lines 1 and 2.

STEP 2. Draw the three-phase compressor, with lead [2] on line 4, the outdoor fan PSC motor, with lead [C] on line 8.

STEP 3. Add the external overloads in L_1, L_2, and L_3 after the fuse.

STEP 4. Place contacts between the fuse and the overload on lines 3, 4 and 5.

STEP 5. The wiring in Step 4 shows control for the compressor only. Wire the outdoor fan motor to start and stop with the compressor by connecting it from L_1, to L_2 on the load side of the contacts and outside the overloads.

STEP 6. On line 11, draw the indoor fan motor and its contacts. Draw the 240-V step-down transformer with a 208-V tap, on lines 13 and 14. Connect all of these elements to L_1, and L_2.

This completes the high-voltage section. Re-check your drawing and then start the low-voltage control circuit.

STEP 7. Protect the low side of the transformer with a fusible link on line 14.

STEP 8. Draw a pair of field-wired terminals [V] on line 14, [C] on line 18, [F] on line 20.

STEP 9. On line 18, draw the compressor contactor coil and connect it to the transformer.

STEP 10. While on line 18. First, put in the high-pressure switch. Now add the three overload contacts OL_1, OL_2, OL_3 and connect all these safety devices to terminal [C].

STEP 11. On line 16, draw in the high impedance relay coil, in series with the contactor coil, (as described in Section II). The normally-closed contacts of the impedance relay are on line 18 and in series with the contactor coil.

STEP 12. On line 20, install the indoor fan relay coil, and complete the wiring from terminal [F].

You have now completed the schematic for the package unit. Go over the steps again until you have a complete understanding of how the low- voltage circuit controls the operation of the system from the thermostat.

The tear-out
schematic layout sheet
can be found on page 241.

STEP 1

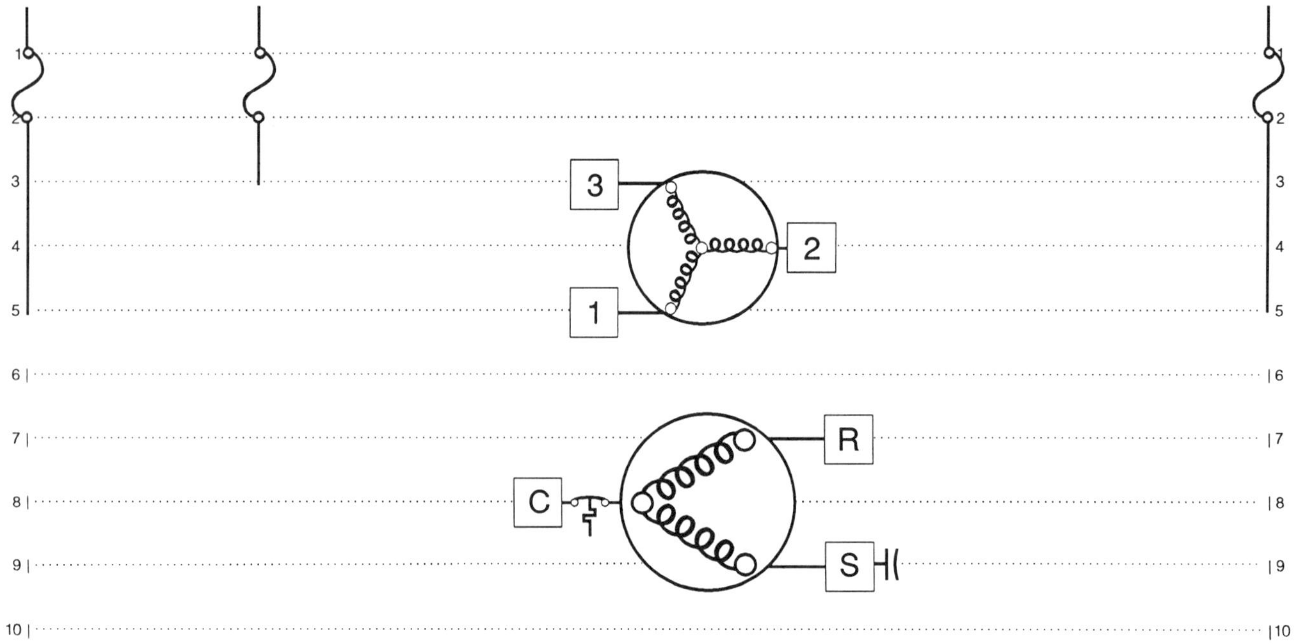

STEP 2

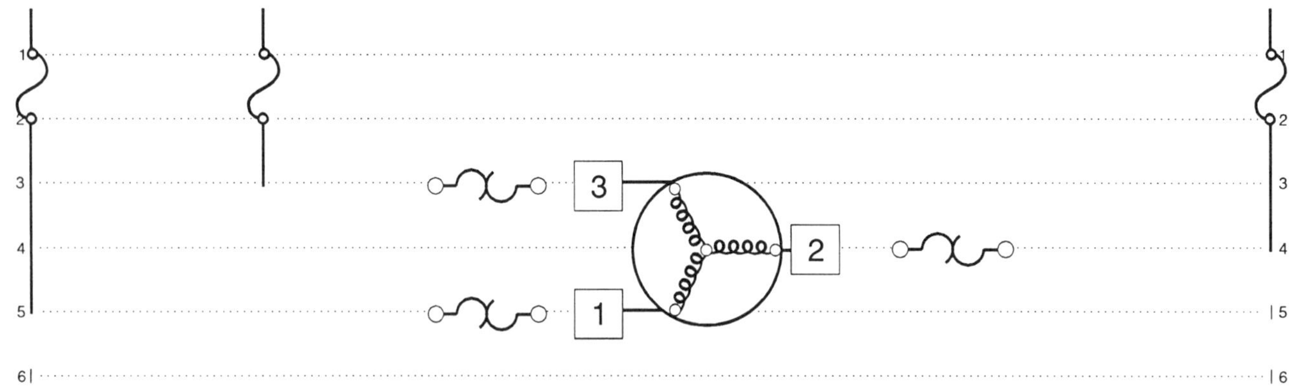

STEP 3

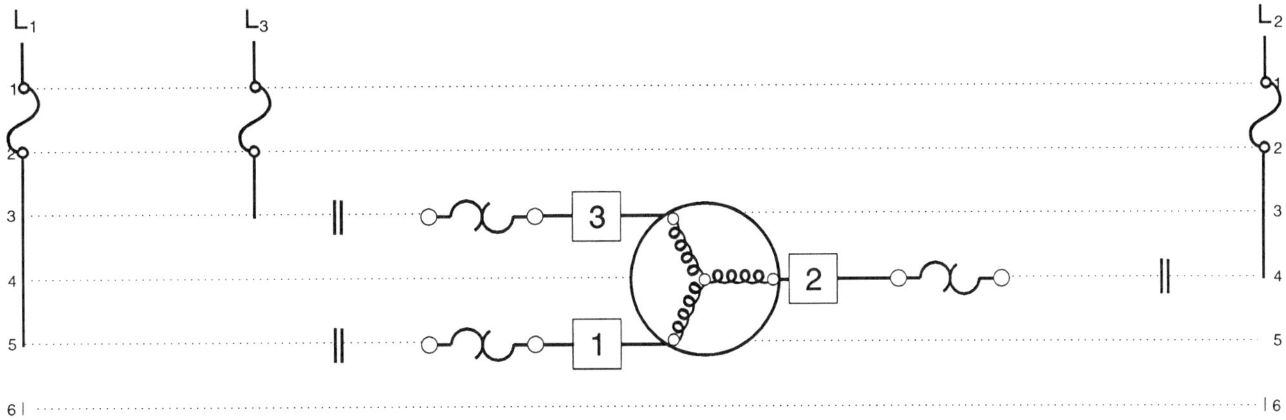

STEP 4

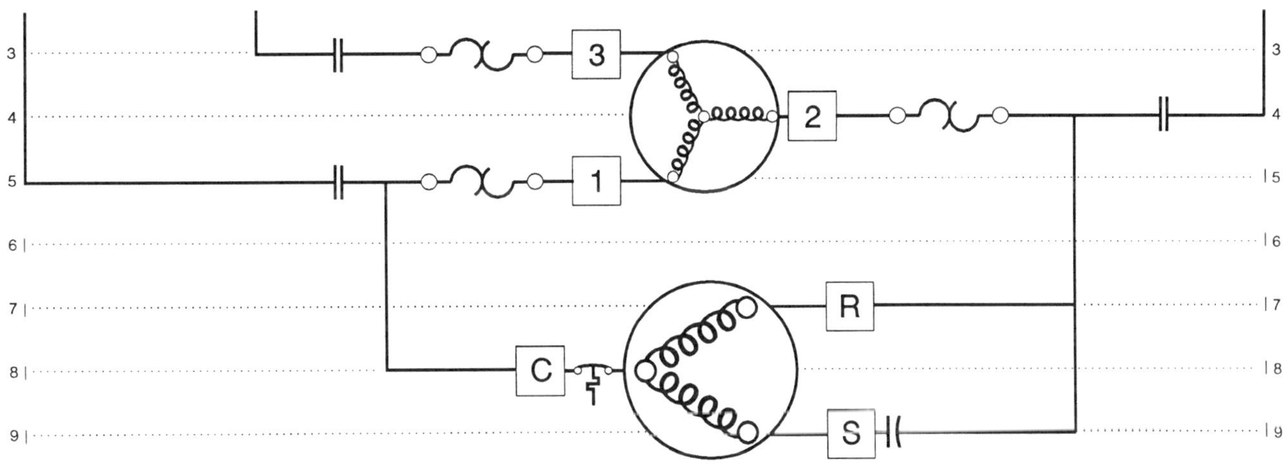

STEP 5

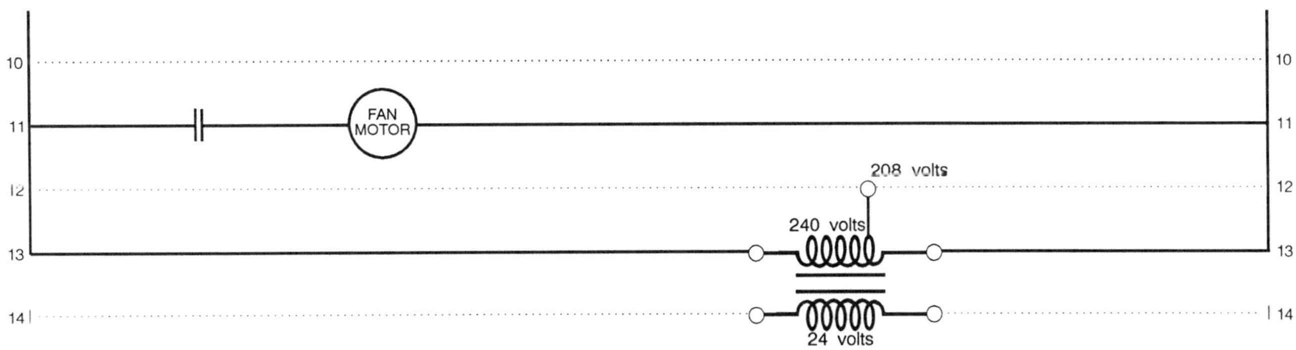

STEP 6

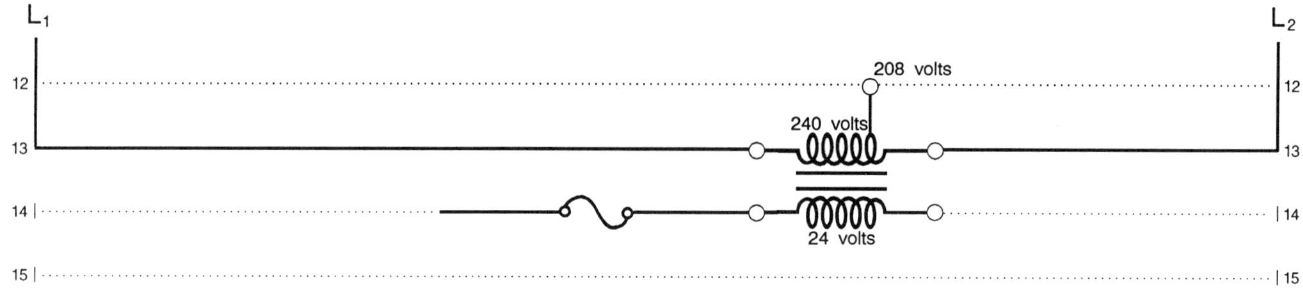

STEP 7

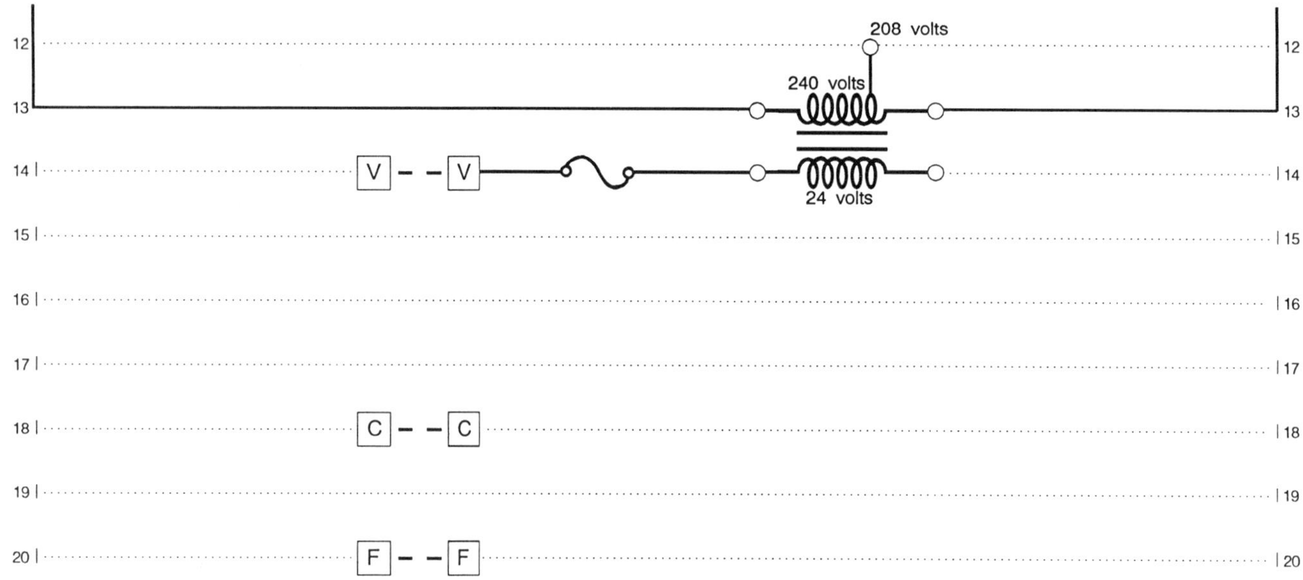

STEP 8

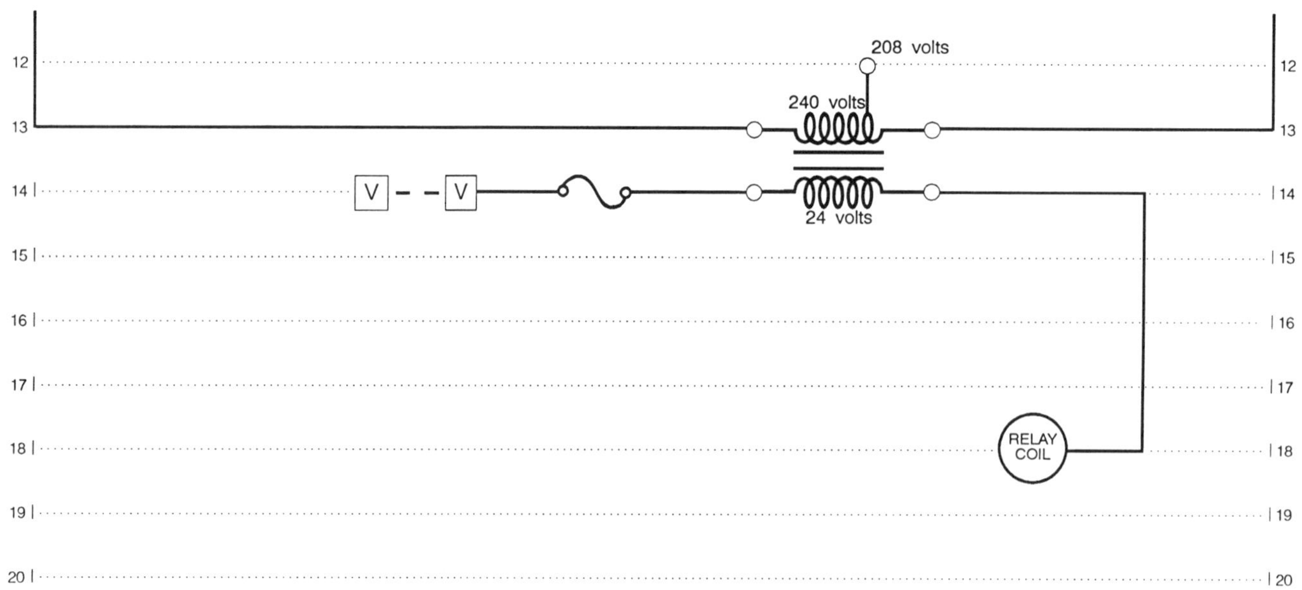

STEP 9

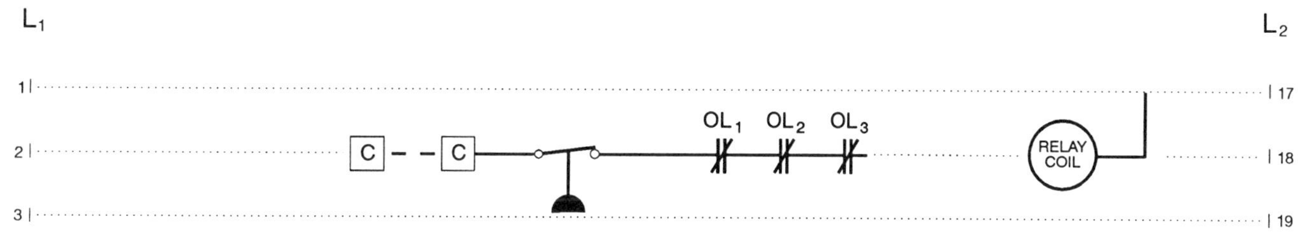

Step 10

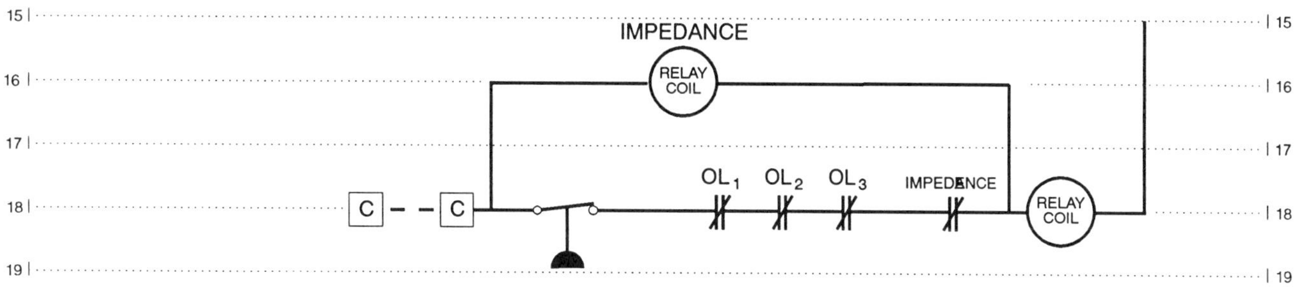

Step 11

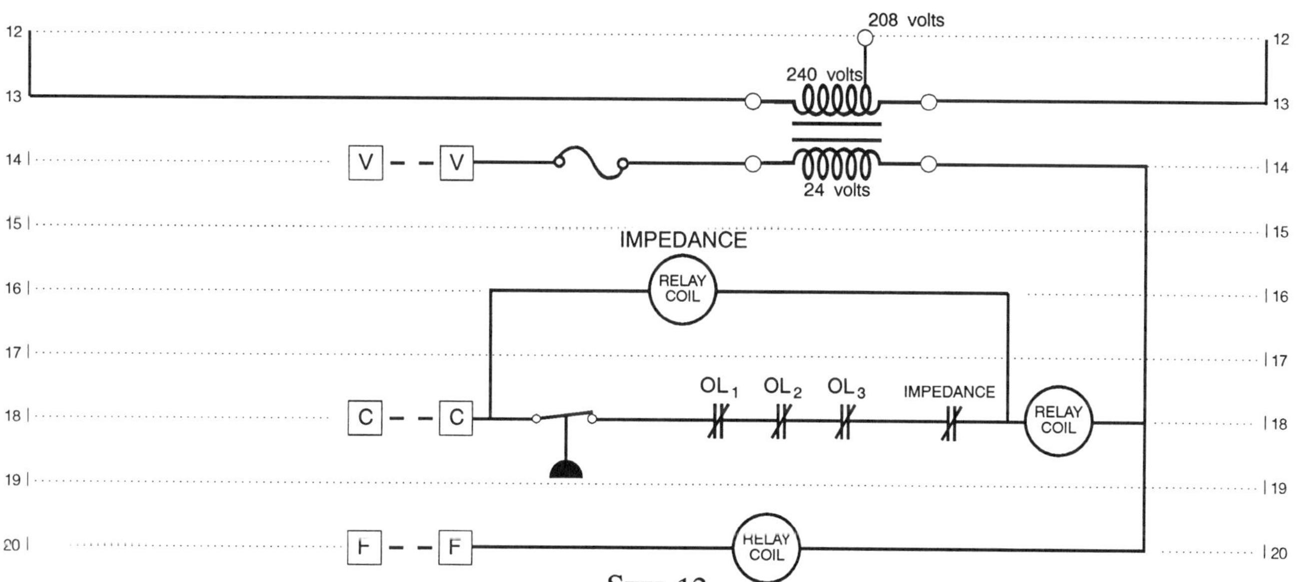

Step 12

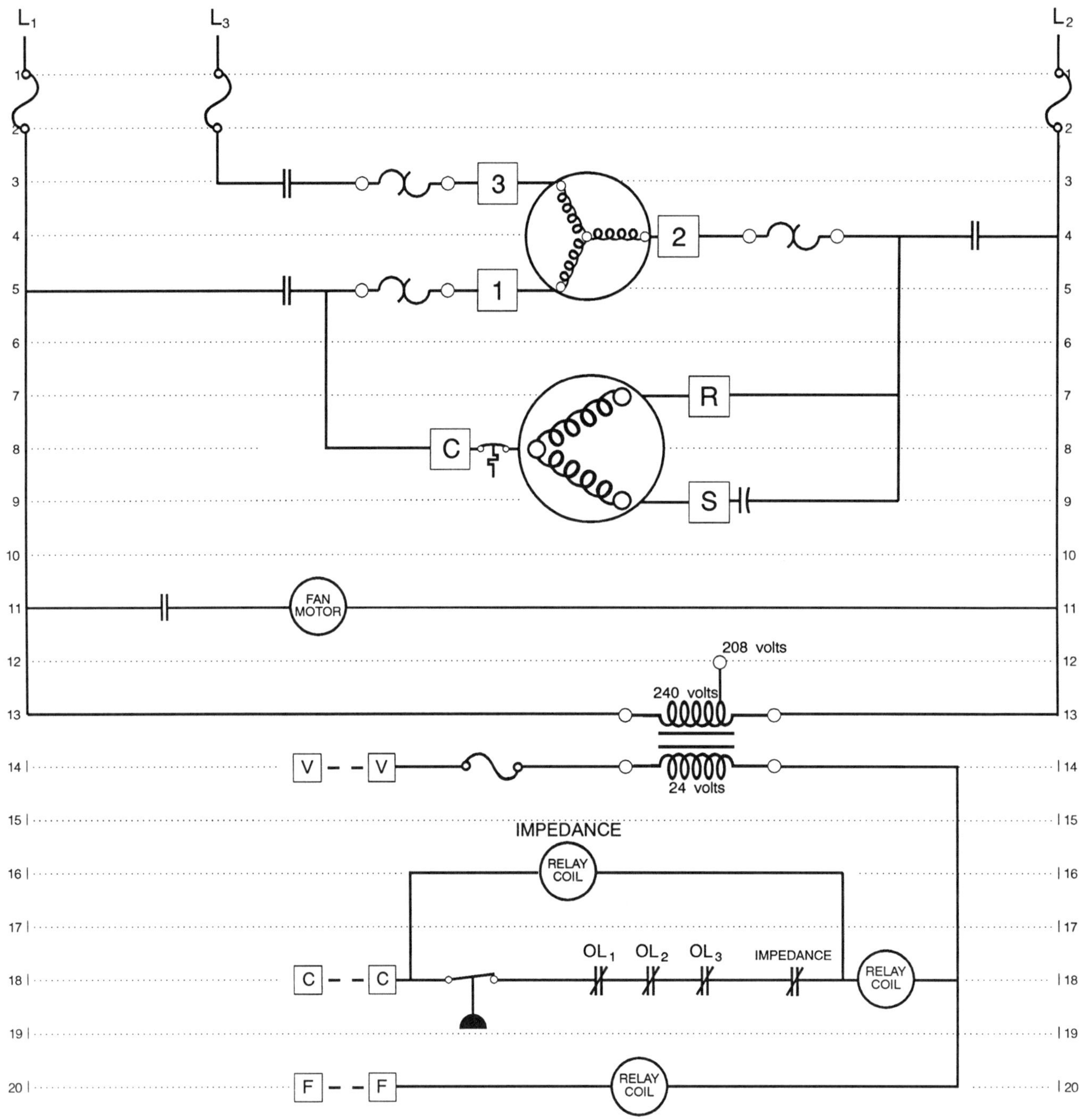
L1
L3
L2
3
2
1
R
C
S
FAN MOTOR
208 volts
240 volts
24 volts
V - - V
IMPEDANCE
RELAY COIL
OL1 OL2 OL3
IMPEDANCE
C - - C
RELAY COIL
F - - F
RELAY COIL

Questions for Example 9

What voltages would you expect to find at the following contacts and loads? Read the questions carefully before answering.

Q-1, Compressor terminals 1 to 3, lines 3 and 5, with the compressor operating?
A-1, ____V

Q-2, Compressor terminals 1 to 2, compressor operating?
A-2, ____V

Q-3, Terminals [R] to [S], lines 7 and 9, on the condenser fan motor with the compressor operating?
A-3, ____V

Q-4, Evaporator fan motor terminals, line 11, with the compressor operating?
A-4, ____V

Q-5, Transformer primary with the thermostat in the OFF position?
A-5, ____V

Q-6, Thermostat terminals [V] to [C], with the thermostat in the OFF position?
A-6, ____V

Q-7, Thermostat terminals [C] to [F], with the thermostat in the OFF position?
A-7, ____V

Q-8, Impedance relay coil with the system operating normally?
A-8, ____V

Q-9, Impedance contacts with the system locked out and all safety switches reset, closed?
A-9, ____V

Q-10, Cooling contactor coil, line 18, with the system locked out?
A-10, ____V

List, for each problem, all of the electrical items that could be checked for defect.

Q-1, The condenser fan motor is not running but the compressor and indoor fan motor are operating.
A-1, ______________________________

Q-2, The condenser fan motor and the indoor fan motor are running but the compressor is off.
A-2, ______________________________

Q-3, The impedance contacts are open and the compressor is locked out.
A-3, ______________________________

Q-4, If OL_3, line 18, opened, would the condenser fan motor continue to run?_______ Would the evaporator fan motor continue to run?_______

Q-5, If fuse L_3 opened would the condenser fan motor continue to run?______ Would the evaporator fan motor continue to run?______

EXAMPLE 10
ELECTRIC HOT-AIR FURNACE

During the past few years, electric resistance hot-air furnaces have been growing in popularity, especially in areas where natural gas is either scarce or expensive.

One of the problems with electric heat is the surge in power demand, which is expensive for utilities. Therefore, as you will see, electric furnaces are designed to bring the power on in steps, rather than in one massive surge.

The unit that we will now examine is supplied as a complete, factory-wired package, lacking only the thermostat.

1. DEFINE THE FUNCTION OF THE EQUIPMENT

This is an automatically-operated, 240-V, electric resistance, hot-air furnace, designed to energize the heating elements in three steps. Due to the high amperage, all high-voltage circuits are doubly fused and all heating elements are isolated by additional fuses. For added safety, each heating stage is equipped with a thermal limit switch that opens if the fan fails or any malfunction leads to overheating.

2. LIST THE ELECTRICAL COMPONENTS

- A 240 V, 60 Hz, single-phase (1ϕ), ac power source is required.
- A fuse is required by the Safety Code.
- A thermostat, with terminals [V] (common), [H] (heat), and [F] (fan), is needed for automatic control.
- A permanent split capacitor motor is required for hot air distribution.
- A fan relay coil and fan relay contacts are needed for fan control. Three pairs of resistance heating elements each separated by a fusible link, are required for heat output.
- Each high voltage circuit must be doubly protected by cartridge fuses.
- Each heating stage is protected by a thermal limit switch.
- Each stage is sequenced by a time-delay relay composed of a heater and two sets of normally-open contacts : one set in the low- voltage circuit, the other in the high- voltage circuit.
- A step-down transformer is used to supply 24-V power to the control circuits.

3. GO STEP-BY-STEP THROUGH THE ELECTRICAL SEQUENCE

- When the thermostat closes on a temperature drop, it completes a circuit from [V] to [H] and [F]. This energizes the indoor fan relay coil and closes the contacts to the fan motor.
- A second circuit, through [H], energizes the 1TR time delay heater. After 20 seconds, the heater closes the 1TR-1 contacts and completes the circuit to the 1HTR first-stage heaters.
- At the same time, a second set of contacts, 1TR-2, close and complete a circuit to the second stage, 2TR, time-delay relay heater.
- After 20 seconds, the second time-delay heater, 2TR, completes a circuit through the 2TR-1 contacts to the second stage heaters 2HTR and, through contacts 2TR-2, to the third stage time delay heater, 3TR.
- After 20 seconds, the third time-delay heater, 3TR, closes the 3TR-1 contacts to the third stage heater, 3HTR, also closing the contacts 3TR-2, on line 2. The contacts on line 2 create a parallel circuit to bypass the fan relay contact on line 3, to keep the fan running until 3HTR is de-energized.

- Should localized heating occur at any one of the heaters, the closest thermal limit switch will open and interrupt the circuit at that stage. If the overheating is widespread, due to fan failure, all three thermal limit switches will open and cut off the heaters, even while the thermostat continues to call for heat.

- If there is a short, or any other source of excess amperage, the fusible links will melt open and isolate the overloaded circuit or component.

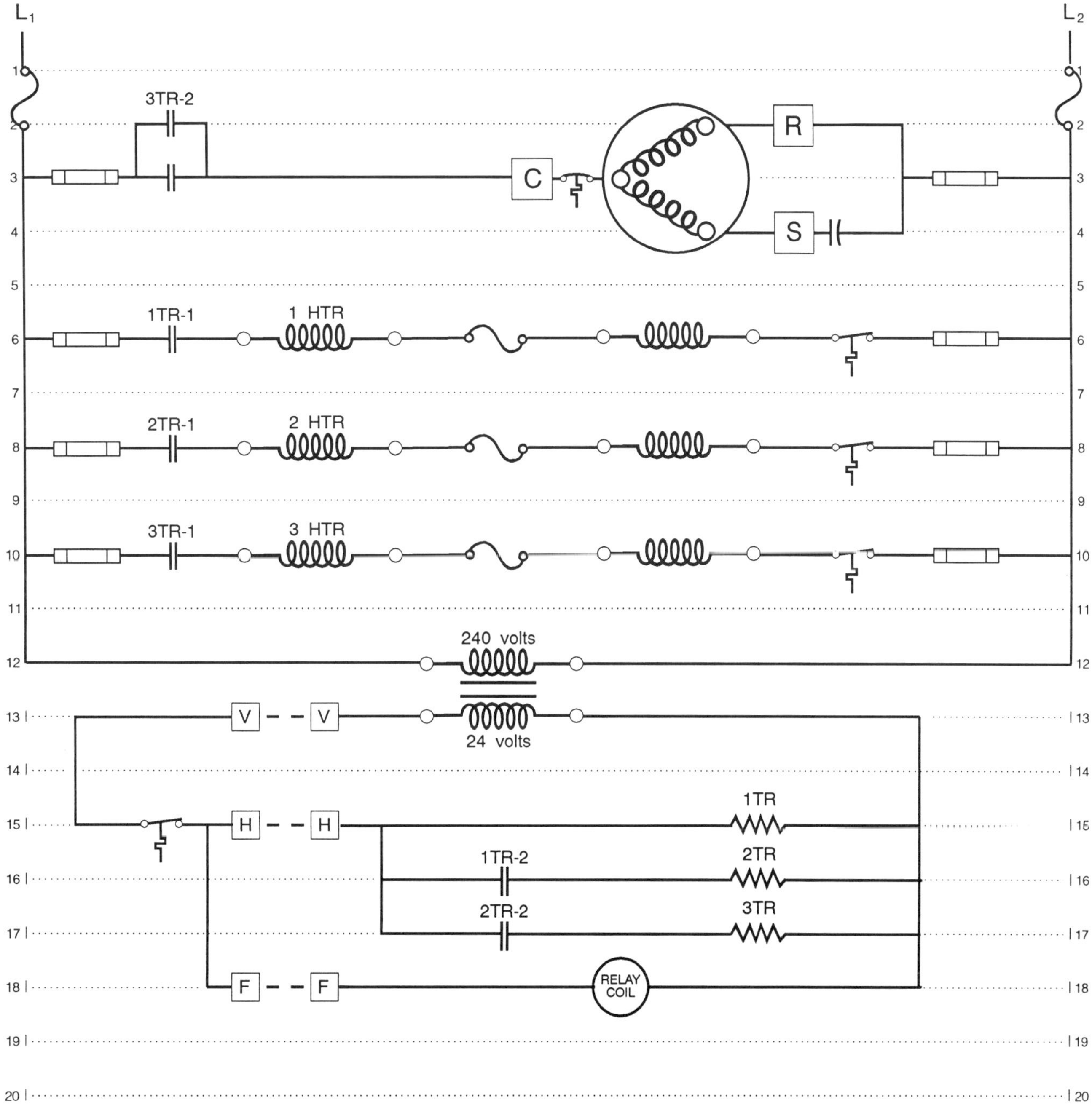

4. Draw the Schematic

STEP 1. Insert a fuse in L_1 and L_2 between lines 1 and 2. Next, insert a set of normally- open contacts in line 2, parallel to the fan contacts in line 3 (drawn in the next step).

STEP 2. On line 3, draw a cartridge fuse, the contacts of the indoor fan relay, and a PSC motor. Join the run [R] and start [S] windings and connect them to a cartridge fuse, which is joined to line L_2 on line 3.

STEP 3. On line 6, draw a cartridge fuse, the contacts (1TR-1) for the 1TR-1 time-delay relay, a heater coil, a fusible link, another heater coil, a thermal limit switch and another cartridge fuse. Connect this array in series with L_1 and L_2

STEP 4. Repeat Step 3 on line 8 and again on line 10, changing the labels on the relay contacts and heater coils to indicate that they are in the second and third heating stages, respectively.

STEP 5. Draw the 240-V side of a step-down transformer on line 12, and the 24-V side on line 13. Connect the 24-V side to terminal [V] on line 13.

STEP 6. On line 18, draw terminal [F] and the indoor fan relay coil.

STEP 7. On line 15, draw terminal [H] and the 1TR heater for the first time delay relay.

STEP 8. On line 16, draw the 1TR-2 contacts (1TR-2) for the first time-delay relay and the 2TR heater for the second time-delay relay.

STEP 9. On line 17, draw the 2TR-2 contacts for the second time-delay relay and the 3TR heater for the third time-delay relay.

STEP 10. Connect the TR heaters and contacts on lines 15,16, and 17 to terminal [H]. Connect all of the low-voltage circuits to the return side of the transformer on line 13.

STEP 11. On line 15 draw a heating thermostat connected to terminal [H]. Connect the other terminal of the thermostat to terminal [V] on line 13. Next draw a line from terminal [F] on line 18, to terminal [H] on line 15. To complete the schmatic, field wire _ _ _ terminals [V] to [V], [H] to [H], and [F] to [F].

The tear-out
schematic layout sheet
can be found on page 243.

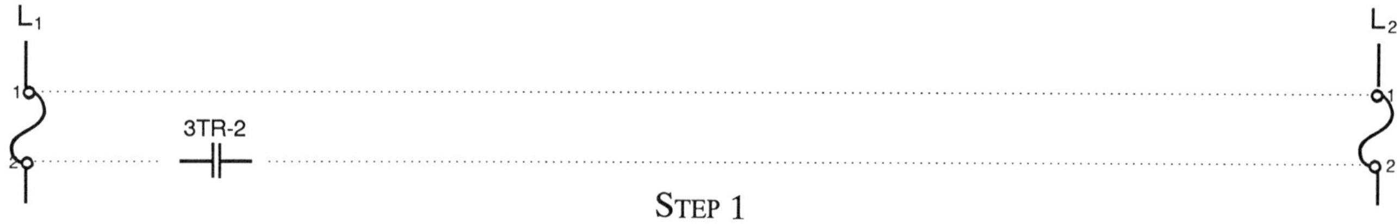

STEP 1

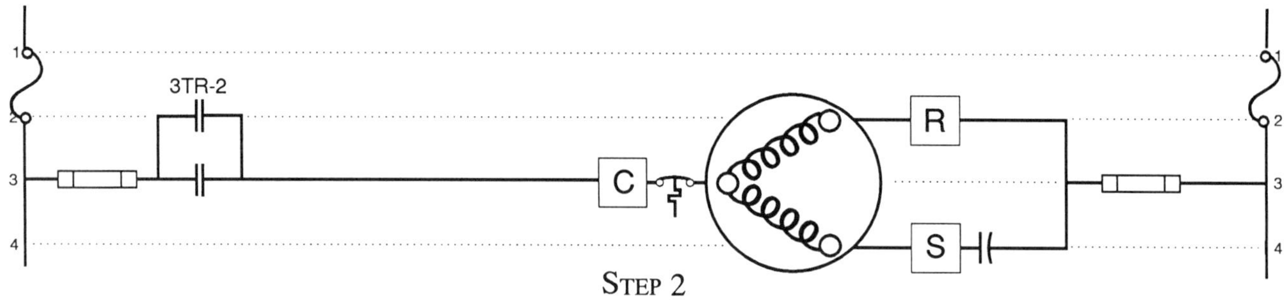

STEP 2

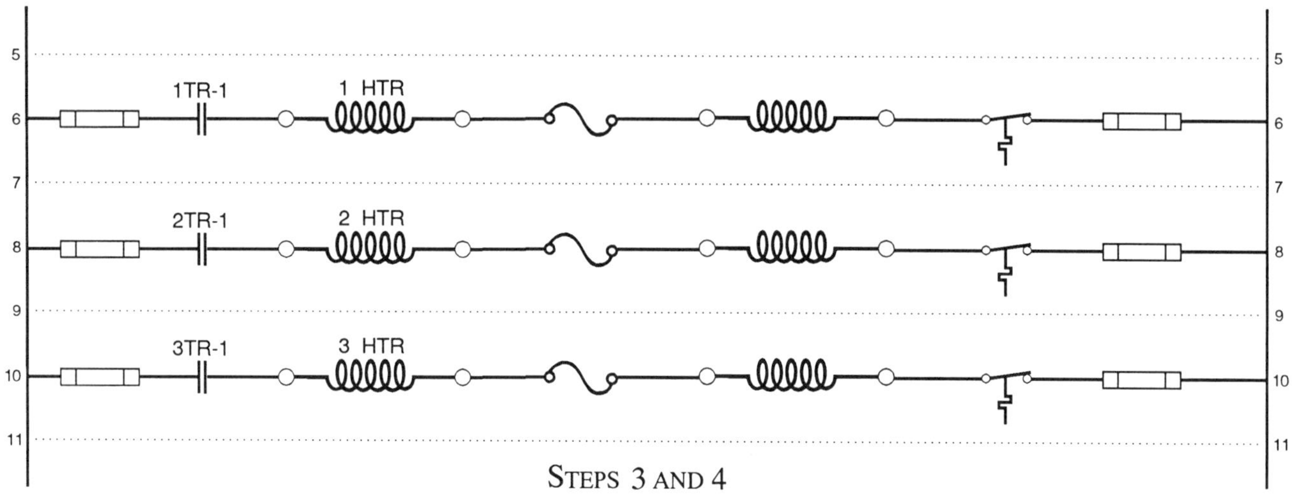

STEPS 3 AND 4

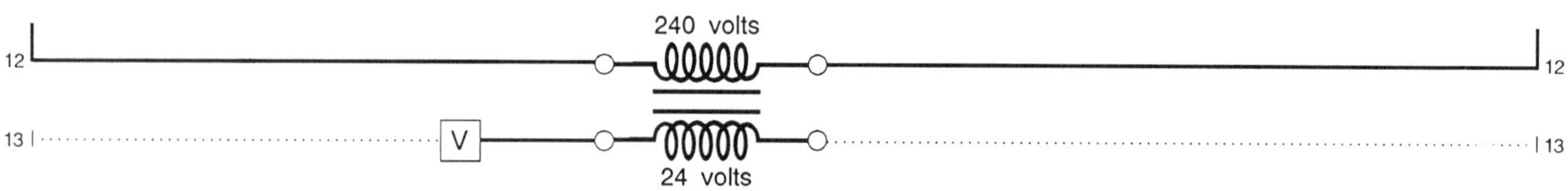

STEP 5

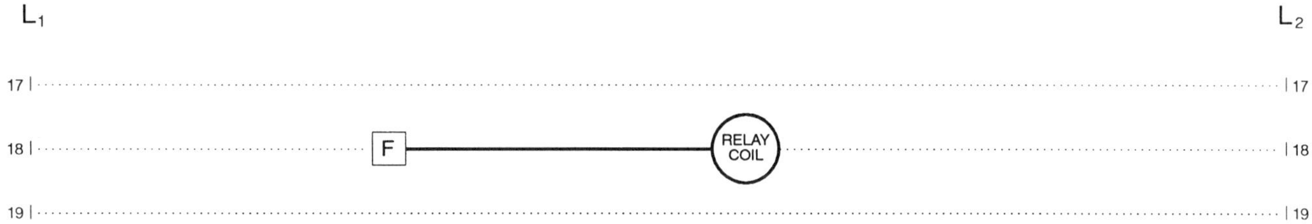

Step 6

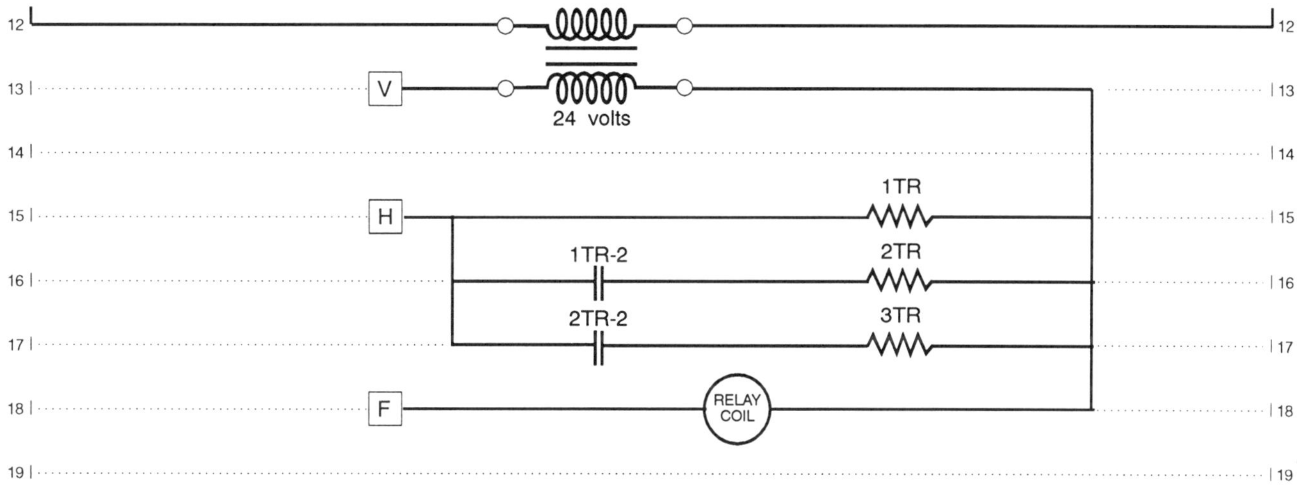

Steps 7, 8, 9 and 10

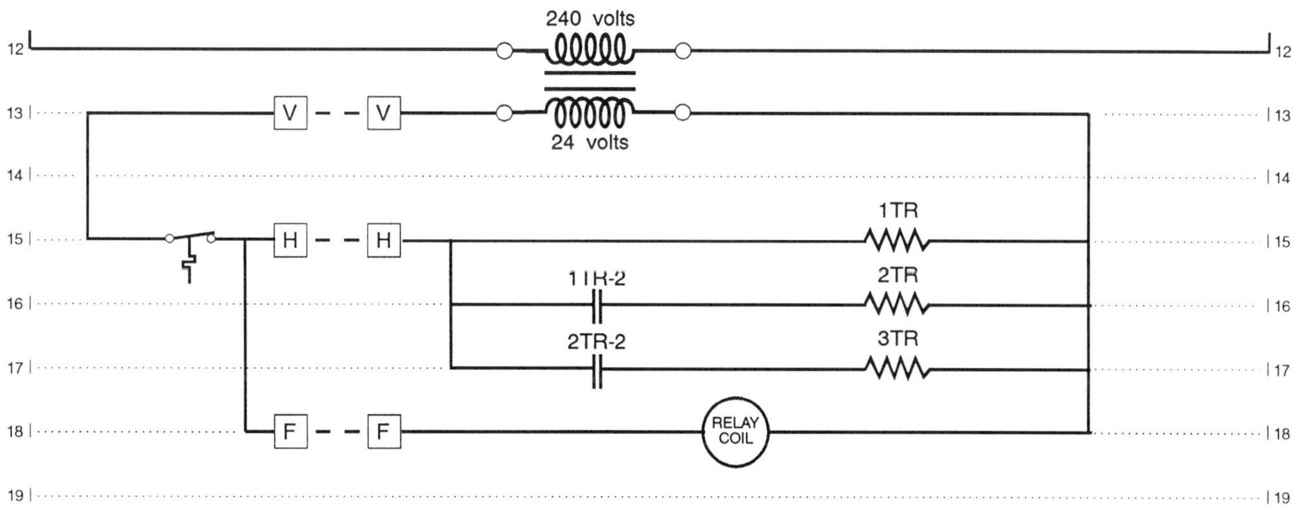

Step 11

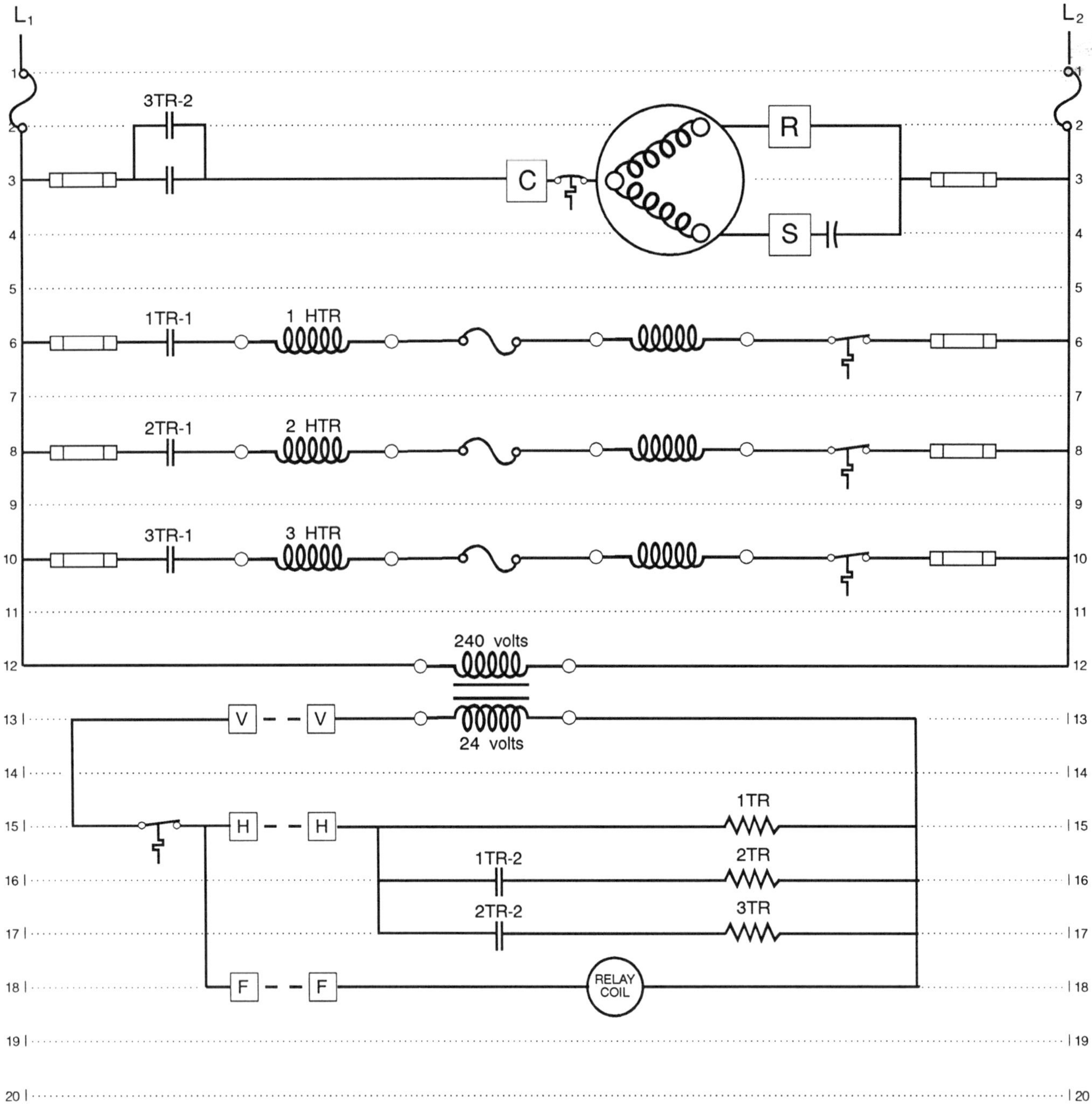
L1
L2
3TR-2
C
R
S
1TR-1
1 HTR
2TR-1
2 HTR
3TR-1
3 HTR
240 volts
V
V
24 volts
H
H
1TR
1TR-2
2TR
2TR-2
3TR
F
F
RELAY COIL

Questions for Example 10

What voltages would you expect to find at the following contacts and loads? Read the questions carefully before answering.

Q-1, Fan motor terminals [R] to [C], lines 2 and 3, with the fan operating?
A-1, ____V

Q-2, Limit switches, lines 6, 8, 10, with the heating system operating?
A-2, ____V

Q-3, Thermostat terminals, [H] to [F], with the thermostat in the OPEN position?
A-3, ____V

Q-4, Terminals [V] to [F] with the thermostat in the OPEN position?
A-4, ____V

Q-5, Time delay relay coil lTR, line 15, with the heat on?
A-5, ____V

Q-6, Contacts 1TR-1, line 6, with the heater 1HTR energized?
A-6, ____V

Q-7, Relay coil, line 18, with the thermostat contacts, line 15, closed?
A-7, ____V

Q-8, Contacts 2TR-2, line 17, with the time-delay coil, line 16, energized?
A-8, ____V

Q-9, With all heaters energized what voltage across contacts 3TR-2, line 2?
A-9, ____V

Q-10, With all heaters energized, what voltage across contacts line 3?
A-10, ____V

List, for each problem, all of the electrical items that could be checked for defect.

Q-1, If lHTR and 3HTR are energized but 2HTR does not energize.
A-1, ______________________________

Q-2, If 1HTR is energized but 2HTR and 3HTR does not energize.
A-2, ______________________________

Q-3, If all three limit switches opened.
A-3, ______________________________

Q-4, The indoor fan motor stopped instantly when the thermostat opened.
A-4, ______________________________

Q-5, Could heater 3HTR energize if the time-delay relay coil 2TR is open? ______ Explain your answer.
A-5, ______________________________

Section VI

MANUFACTURERS' SCHEMATICS

THE WIRING DIAGRAM

A wiring or connection diagram, sometimes called a pictorial, accompanies most schematic drawings. Unlike the schematic, the wiring diagram indicates the general position of each electrical device, such as relays, disconnects, motors, transformers, switch contacts, capacitors, etc. Some wiring diagrams indicate the color of each wire and some show the identifying numbers appearing on the wires.

Figure 6-1 is a typical wiring diagram. It indicates the boundary of the control panel, and all the devices within the panel are shown in their relative positions. External items are shown outside the panel boundary with some indication of their location relative to the panel. In some cases, related items cannot be easily shown on the diagram. When this happens, the manufacturer usually notes, on the diagram, where the unshown device is located.

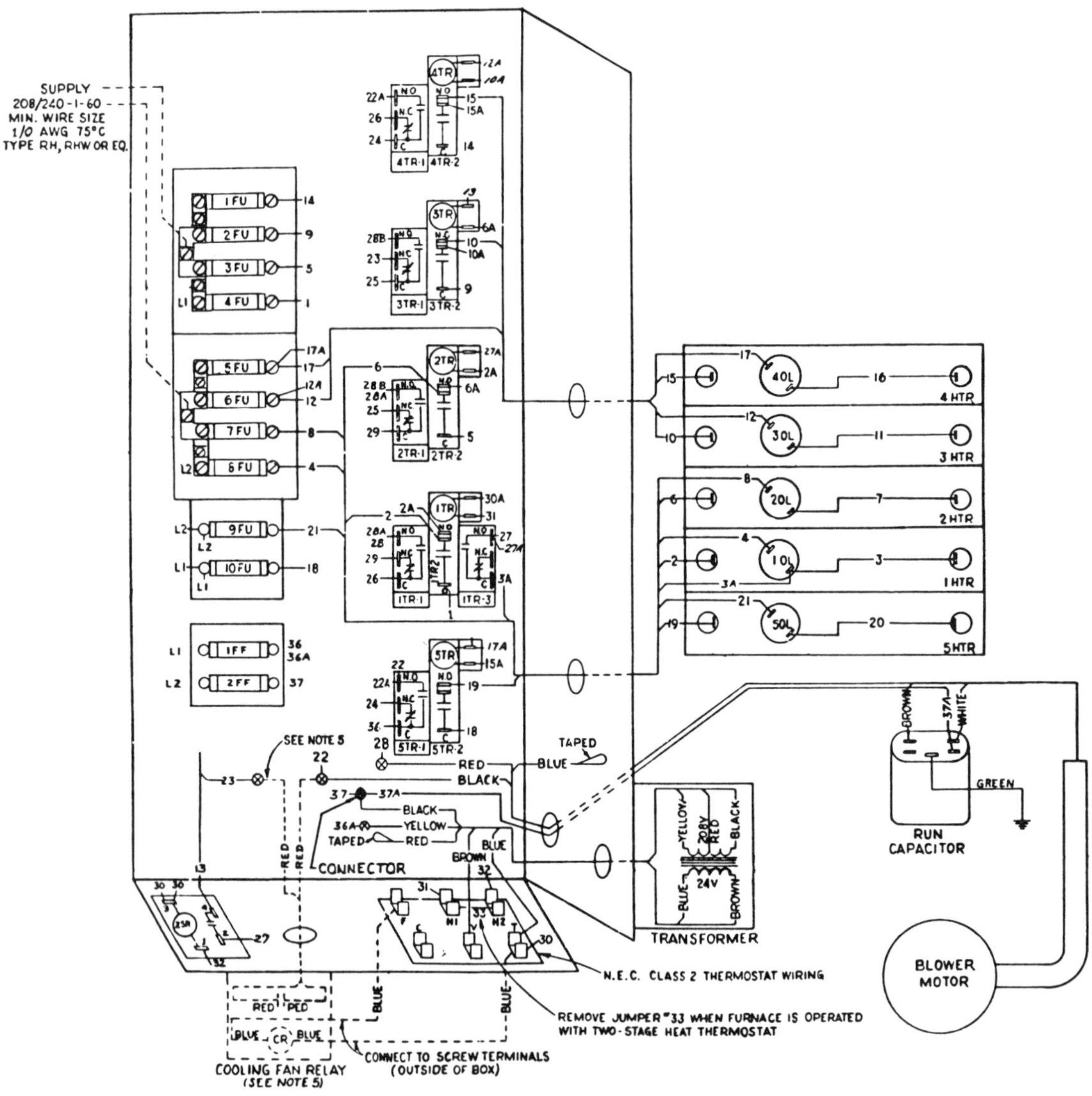

Figure 6-1. Courtesy, *York Division, Borg-Warner Corporation.*

Wiring on the connection diagram is shown in a number of ways. The three most popular methods are:

1. Each wire is drawn from terminal to terminal. A good illustration of this method is the wiring diagram for the Frigidaire side-by-side refrigerator/freezer, including the optional ice maker and butter conditioner, Figure 6-2. This wiring diagram shows every wire and electrical device in a pictorial fashion.

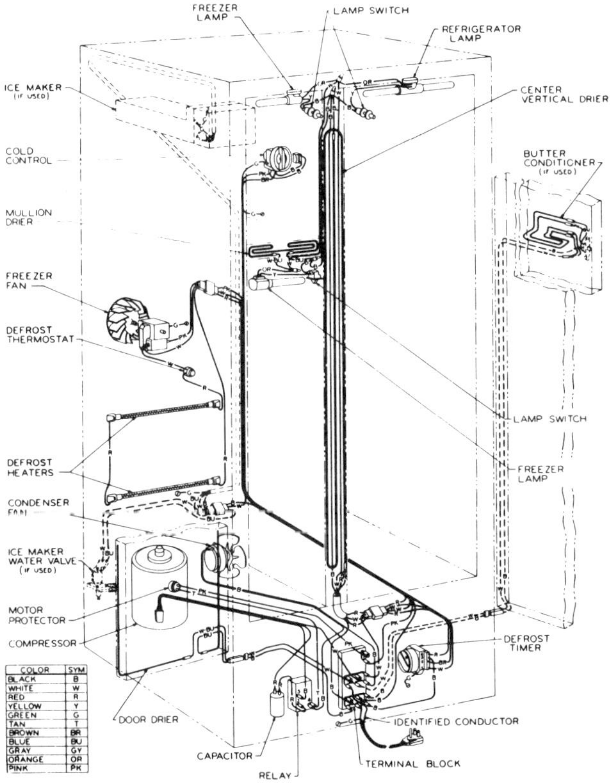

Figure 6-2. Courtesy, *Frigidaire Division*.

2. Groups of wires are shown as a harness, with single wires connected at each component. York uses this method very effectively, Figure 6-3, in routing the four wires from the FAN MOTOR. York also prints identifying numbers on each wire to make circuit tracing much easier.
   The broken lines — — —show field wiring necessary to install the capacitor and potential relay.

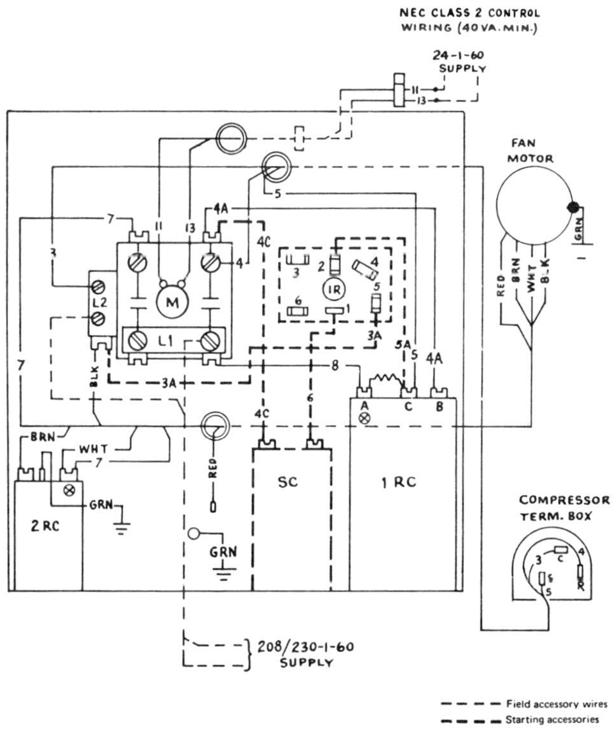

Figure 6-3. Courtesy, *York Division, Borg-Warner Corporation.*

3. Little or no wiring is shown from terminal to terminal. Figure 6-4 is typical of this method which is being adopted by more manufacturers. Note that each terminal is assigned a number. You can conclude that each pair of terminals having matching numbers is joined by wiring. Identical numbers are used on the schematic, making it easy to move from the schematic to the pictorial drawing when tracing a circuit.

   This method produces a cleaner drawing that is uncluttered by a maze of wiring. However, it may prove difficult for beginners who are accustomed to tracing a line from point to point. Once mastered, this style proves to be much easier to use. With no wires to follow, your eyes are not distracted. Very often when a group of wires is drawn close together, your eyes skip to an adjacent wire that misleads you to the wrong component.

Connection diagrams will usually show the coil and contacts that are within the components, with the contacts shown in the normally-open or normally-closed position.

Remember, the wiring, or pictorial, diagram is used to locate the defective component . . . after the potential problem has been isolated on the schematic diagram.

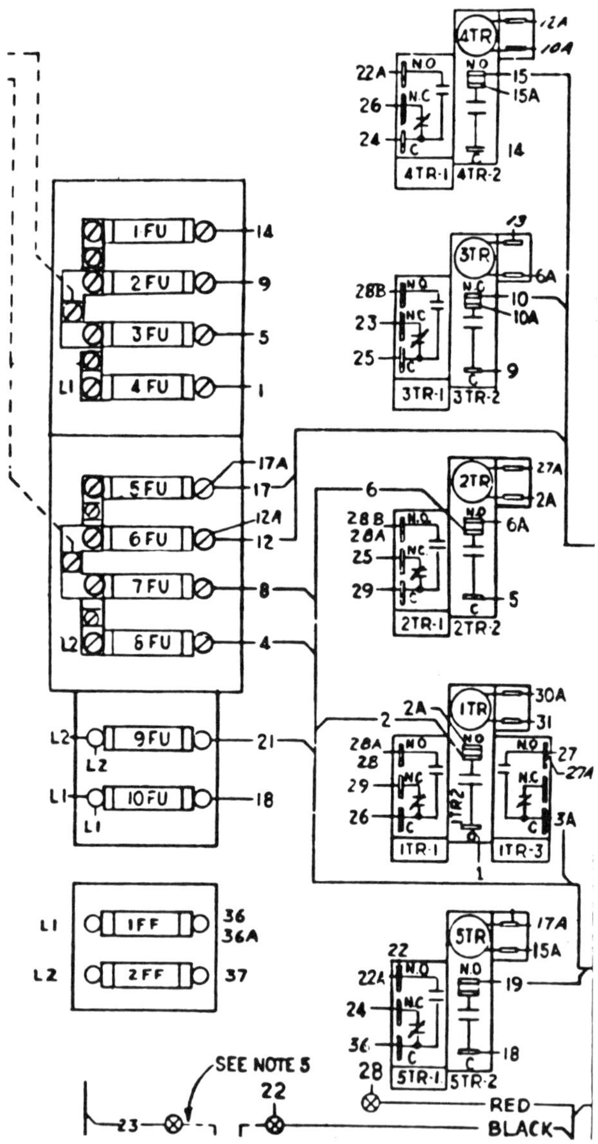

Figure 6-4. Courtesy, *York Division, Borg-Warner Corporation.*

TROUBLESHOOTING THE ELECTRICAL CIRCUIT

The electrical schematics shown in the following examples are for heating, cooling and refrigerating units that are in use today. They were selected to demonstrate the many different styles used by manufacturers in their schematics.

Before you proceed, you must have a system or plan for using the schematic to isolate the electrical breakdown in the circuit. Without a system, it is easy to overlook or forget a less obvious point and misdiagnose the problem.

Here is a recommended procedure that has been very effective over the years. There are five steps:

STEP 1. Read the schematic to determine the electrical sequence, or *what happens next*. This is the most important step, since it is almost impossible to troubleshoot any equipment without a prior knowledge of how it operates. It is similar to troubleshooting a moon rocket without knowing the location of the start button.

STEP 2. Check the primary voltage to the equipment. It is not enough to say that there is voltage to the equipment, the question is, how much voltage?

Many problems are no fault of the equipment, rather they are due to an undetected voltage drop. Low voltage may be caused by undcrsize wire or overloaded transmission lines (the latter can occur during summer's peak air-conditioning hours).

Do not use a light or neon tester to check voltage. They can only tell you that voltage is present. A voltmeter is needed to tell you exactly how much voltage is available.

STEP 3. Check the fuses. It is safer and more accurate if the fuses are removed and checked with an ohmmeter or continuity tester.

STEP 4. Locate the device that does not function, whether it be a motor, solenoid, relay, etc.

STEP 5. Starting at the non-operating device with the power on, use a voltmeter to check power flow through all contacts leading to the non-operative device.

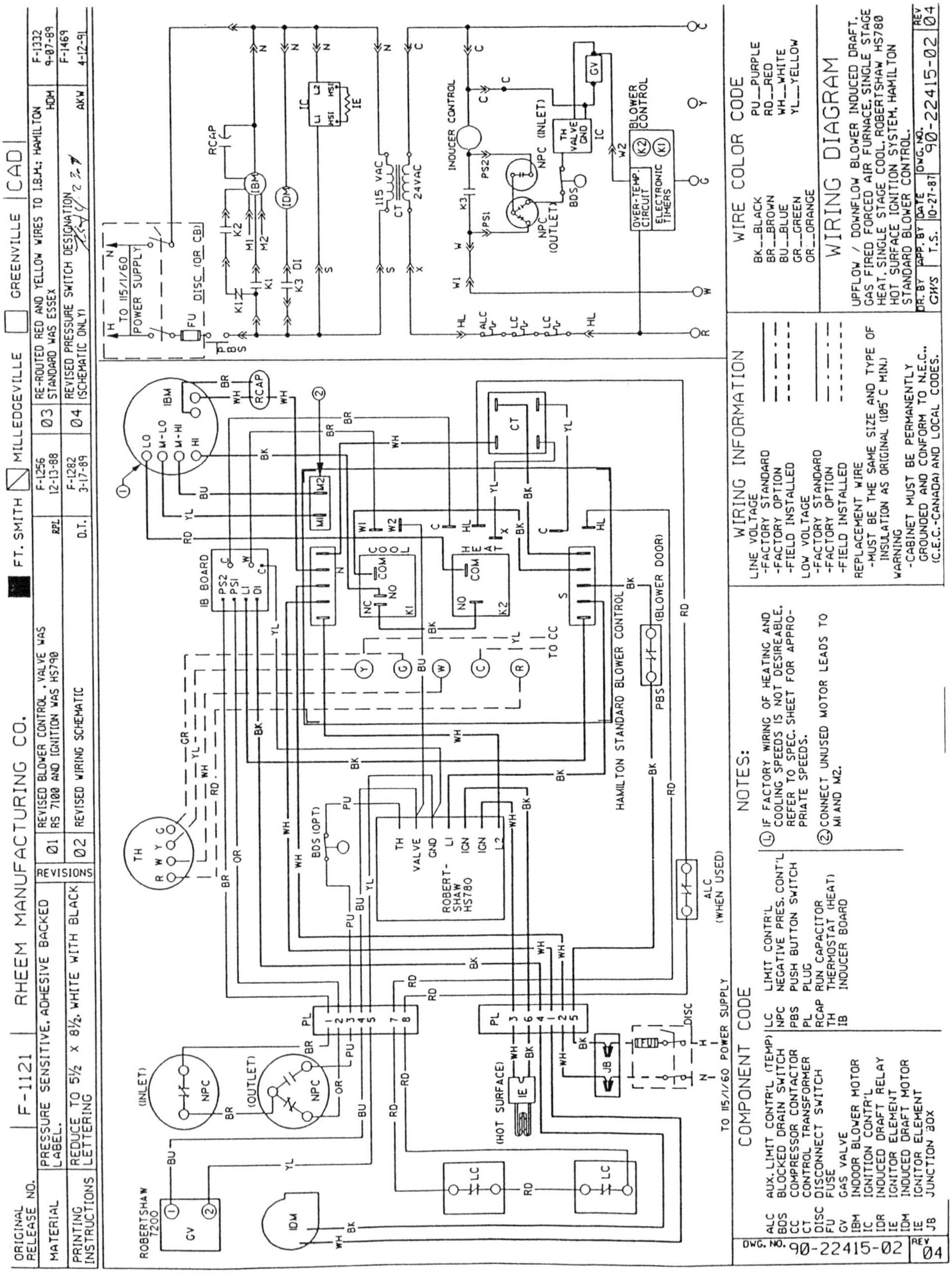
RHEEM MANUFACTURING CO.
F-1121
FT. SMITH
MILLEDGEVILLE
GREENVILLE
CAD
ORIGINAL RELEASE NO.
MATERIAL
PRESSURE SENSITIVE, ADHESIVE BACKED LABEL.
PRINTING INSTRUCTIONS
REDUCE TO 5½ x 8½, WHITE WITH BLACK LETTERING
REVISIONS
01 REVISED BLOWER CONTROL, VALVE WAS RS 7100 AND IGNITION WAS HS790
02 REVISED WIRING SCHEMATIC
03 RE-ROUTED RED AND YELLOW WIRES TO I.B.M.; HAMILTON STANDARD WAS ESSEX
04 REVISED PRESSURE SWITCH DESIGNATION (SCHEMATIC ONLY)
TO 115/1/60 POWER SUPPLY
DISC (OR CB)
INDUCER CONTROL
115 VAC
24VAC
NPC (INLET)
NPC (OUTLET)
BLOWER CONTROL
OVER-TEMP. CIRCUIT
ELECTRONIC TIMERS
IB BOARD
ROBERT-SHAW HS780
BDS (OPT)
HAMILTON STANDARD BLOWER CONTROL
(BLOWER DOOR)
ALC (WHEN USED)
ROBERTSHAW 7200
(HOT SURFACE)
TO 115/1/60 POWER SUPPLY
WIRE COLOR CODE
BK__BLACK
BR__BROWN
BU__BLUE
GR__GREEN
OR__ORANGE
PU__PURPLE
RD__RED
WH__WHITE
YL__YELLOW
WIRING DIAGRAM
UPFLOW / DOWNFLOW BLOWER INDUCED DRAFT. GAS FIRED FORCED AIR FURNACE, SINGLE STAGE HEAT, SINGLE STAGE COOL. ROBERTSHAW HS780 HOT SURFACE IGNITION SYSTEM, HAMILTON STANDARD BLOWER CONTROL.
DWG. NO. 90-22415-02
REV 04
WIRING INFORMATION
LINE VOLTAGE
-FACTORY STANDARD
-FACTORY OPTION
-FIELD INSTALLED
LOW VOLTAGE
-FACTORY STANDARD
-FACTORY OPTION
-FIELD INSTALLED
REPLACEMENT WIRE
-MUST BE THE SAME SIZE AND TYPE OF INSULATION AS ORIGINAL (105° C MIN.)
WARNING
-CABINET MUST BE PERMANENTLY GROUNDED AND CONFORM TO N.E.C., (C.E.C.-CANADA) AND LOCAL CODES.
NOTES:
1 IF FACTORY WIRING OF HEATING AND COOLING SPEEDS IS NOT DESIREABLE, REFER TO SPEC. SHEET FOR APPROPRIATE SPEEDS.
2 CONNECT UNUSED MOTOR LEADS TO M1 AND M2.
COMPONENT CODE
ALC AUX. LIMIT CONTR'L (TEMP)
BDS BLOCKED DRAIN SWITCH
CC COMPRESSOR CONTACTOR
CT CONTROL TRANSFORMER
DISC DISCONNECT SWITCH
FU FUSE
GV GAS VALVE
IBM INDOOR BLOWER MOTOR
IC IGNITION CONTR'L
IDR INDUCED DRAFT RELAY
IE IGNITOR ELEMENT
IDM INDUCED DRAFT MOTOR
JB JUNCTION BOX
LC LIMIT CONTR'L
NPC NEGATIVE PRES. CONT'L
PBS PUSH BUTTON SWITCH
PL PLUG
RCAP RUN CAPACITOR
TH THERMOSTAT (HEAT)
IB INDUCER BOARD

RUUD 90 PLUS GAS WARM AIR CONDENSING FURNACE

The modern condensing furnace presents the technician with a few new service and installation problems. The new conditions that must be considered are the low flue gas temperatures and the water condensation caused by the low flue gas temperatures.

This new generation of heating also introduces many solid-state electronic controls for both safety and user convenience. The RUUD model, 90 PLUS, is a good example of a 100,000 Btu gas-fired hot air heater with solid-state controls. The factory schematic is straightforward and not too difficult to follow, Figure 6-5. To remove the low temperature flue gas, a draft inducer blower and motor is used to force oxygen into the fire and blow the products of combustion through a horizontal plastic pipe to the outdoors. The traditional vertical chimney, so familiar in homes for many years, cannot be used, because the flue gas is not hot enough to rise up through the chimney. To eliminate the energy- using standing pilot flame, a device called a Hot Surface Ignition System is employed. This device will first light the pilot when the thermostat calls for heat and then act as a pilot safety after the pilot flame is ignited. An adjustable timer causes the low-speed blower motor to turn on and off.

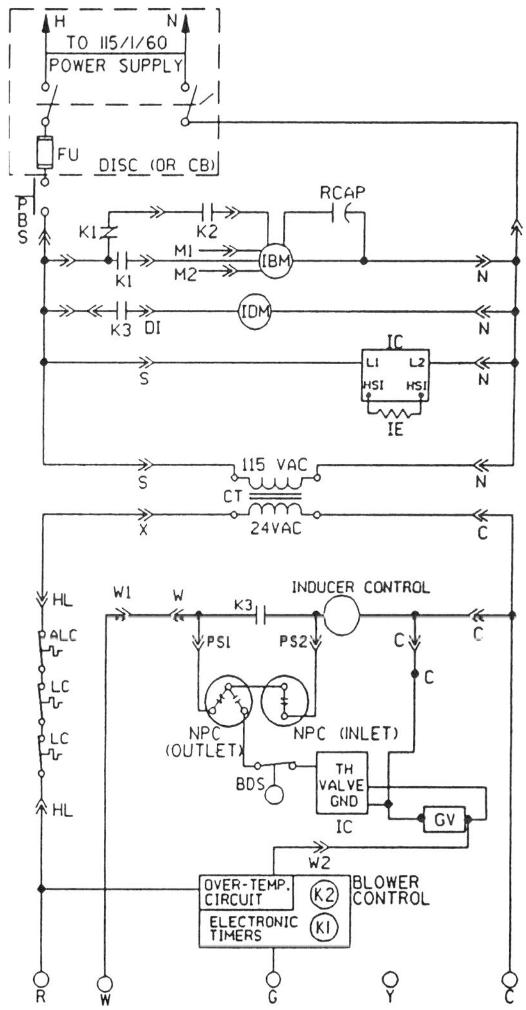

Figure 6-5. Courtesy, *Ruud Corporation*

1. GO STEP-BY-STEP THROUGH THE ELECTRICAL SEQUENCE

To follow the sequence, first set the room thermostat to HEAT, the fan switch to AUTO, and raise the thermostat setting above the room temperature, closing terminals R to W.

- Blower compartment must be closed to close the normally-open door switch PBS, which completes the circuit to the control transformer primary.
- 24 V is available at terminals R and C if the auxiliary limit control (ALC) and the two limit controls (LC) are closed.
- With all controls closed from R to W, a circuit is completed to terminal PS1 on the Inducer Board (IB).
- The circuit continues through the normally-closed contacts in the negative pressure control inlet (NPC) to the normally-closed contacts in the negative pressure control outlet (NPC).
- This circuit is used to by pass the normally-open inducer contacts (K3) and energizing the inducer control, which is part of the inducer board (IB).
- With the inducer draft relay (IDR) energized (the IDR relay is part of the IB board), the normally-open contacts in the 24-V circuit marked K3 will close, paralleling the NPC circuit to the inducer control. At the same time, the normally-open K3 contacts in the 115-V circuit will close, energizing induced draft motor (IDM).
- The negative pressure created by the inlet of inducer draft motor will open the normally-closed contacts and close the normally-open contacts in the NPC (outlet).
- The NC contacts in the NPC inlet control will only open if a negative pressure is created in the burner chamber. This will cause a shut down of the gas valve.
- The circuit to the Ignition Control (IC), through the blocked drain switch BDS, is now completed through terminal P52 from terminal W.
- The Ignition Control circuit will energize the hot surface igniter, in the 115-V circuit.
- The hot surface igniter (IE) is designed to glow bright red hot to ignite the pilot flame.
- If the pilot flame is not proved in a period of 34 seconds, the IE will shut down the main gas valve (GV).
- A circuit is also established to the blower control. After a time delay, the relay K2 will be energized ,starting the low-speed fan motor.

OBSERVING THE FURNACE SCHEMATIC

- The Robert Shaw HS780 controls the hot surface ignitor, gas valve and pilot safety.
- The Hamilton Standard Blower Control times the starting and stopping of both the high-speed fan motor for cooling and the low-speed fan motor for heating.
- The hot surface igniter is used to ignite the pilot flame, after which a pilot safety sensor is energized to protect against pilot flame failure.
- The inducer control board is a part added to the Hamilton Standard Blower Control.

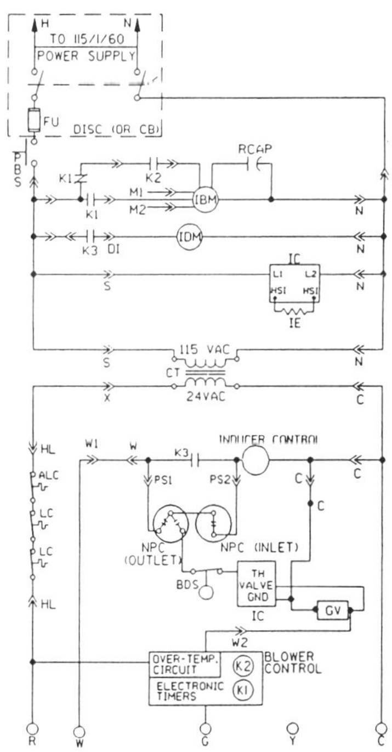

Courtesy, *Ruud Corporation*

Raise the thermostat above room temperature to close contacts R to W. With the heater operating and using a voltmeter, what voltages would you expect to read, with one probe on each side of the following motors, coils and contacts:

1. L_1 to L_2 on the Ignition Control Board
2. Draft Motor (IDM) (IDM)
3. Limit ALC
4. Contacts K2 ⊣⊢
5. Contacts NO K1 ⊣⊢ and NC K1 ⊣/⊢
6. Transformer primary 115 FU
7. Contacts K3 ⊣⊢
8. Gas valve contacts GV
9. Blocked Drain Switch BDS
10. Blower Motor low speed to R

Using The Five-Step Procedure, Solve these problems.

Problem 1. Induced Draft Motor does not come on.

Problem 2. Igniter element does not come on.

Problem 3. Gas valve will not open.

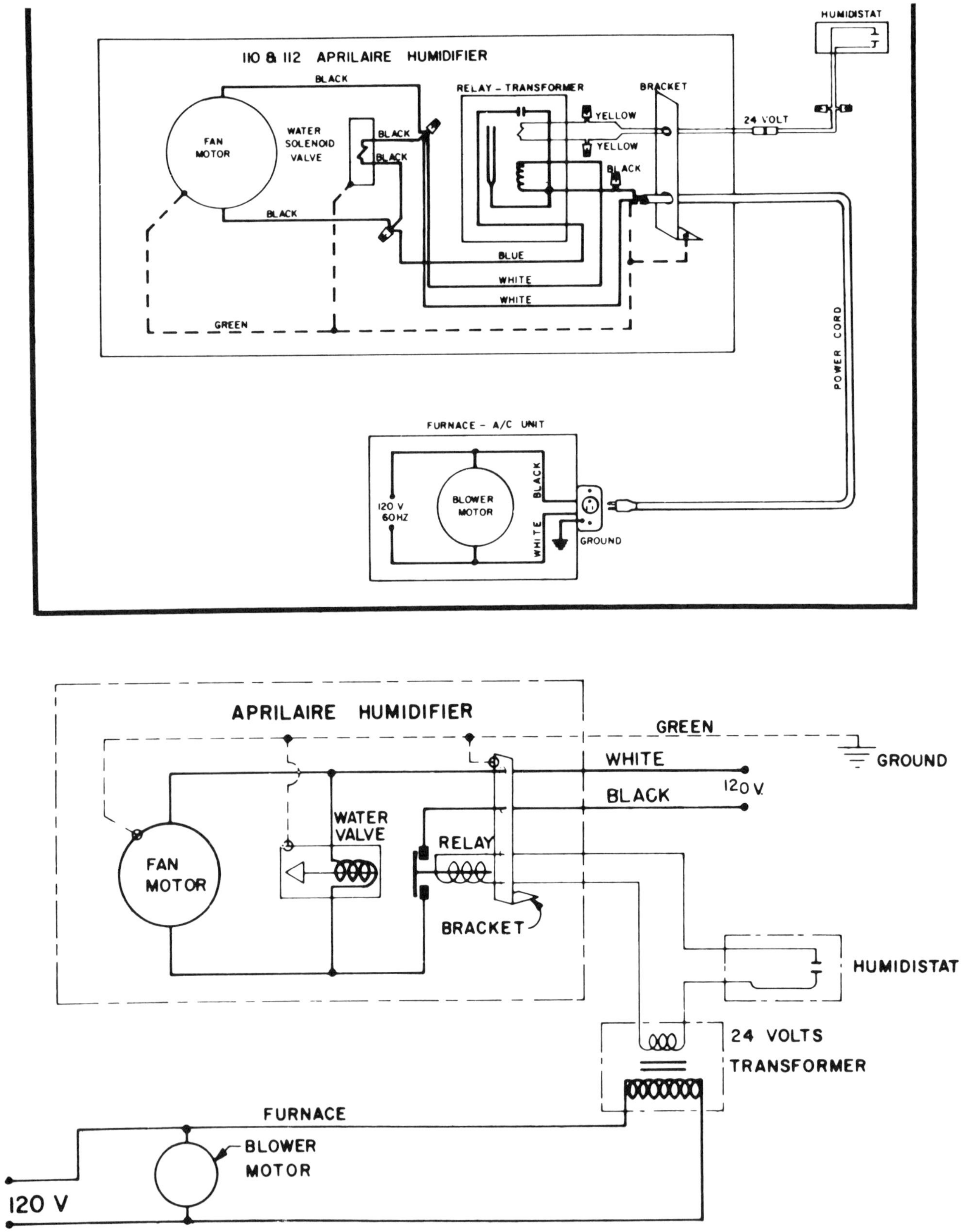

Figure 6-6. Courtesy, *Aprilaire*

APRILAIRE MODEL 110 POWER HUMIDIFIER

During the heating season, warmed indoor air is continuously lost through cracks around windows and doors and replaced by cold outdoor air. When this cold outdoor air is heated, it expands, lowering the relative humidity and causing the air to be dry. The humidifier is used to restore needed moisture to the air to maintain a comfortable humidity level.

All humidifiers are dependent upon heat for evaporation. Humidifiers normally extract this heat from the hot air in the supply plenum at the top of the furnace.

The Aprilaire schematic is a good example of a small humidifier with a 24-V control circuit and a 115-V power circuit ——, Figure 6-6.

The control circuit draws electricity from the furnace blower motor 115-V circuit to ensure that the humidifier only operates when the furnace blower is on. 115 V, from the furnace blower circuit, is applied to the primary of the step-down transformer, which reduces the voltage to 24 V.

The humidistat and coil of the electromagnetic relay are in the 24-V circuit. The power circuit draws electricity from any standard 115 V, 60 Hz supply.

Now, following the five-step troubleshooting procedure:

1. GO STEP-BY-STEP THROUGH THE ELECTRICAL SEQUENCE

- Set the fan control switch on the thermostat to AUTO. This will complete the circuit to the blower motor and simultaneously supply 115 V to the transformer primary.
- Raise the setting on the humidistat to close the humidistat contacts.
- With the humidistat contacts closed, a 24-V circuit is completed from the low-voltage side of the transformer to the relay coil.
- The energized magnetic relay coil will close the relay contacts, completing two parallel 115-V circuits. One 115-V circuit will energize the water valve solenoid, opening the valve to feed water to the humidifier. The other 115-V circuit will start the fan motor to circulate warm furnace air through the humidifier.

By examining the schematic, it can now be seen that...

- The humidifier is totally inoperative when the furnace blower is off.
- If the fan motor fails, the water valve can still operate.
- If the water valve solenoid fails, the fan motor can still operate.
- A failure of the electromagnetic relay coil, humidistat, or transformer will make the humidifier inoperative.

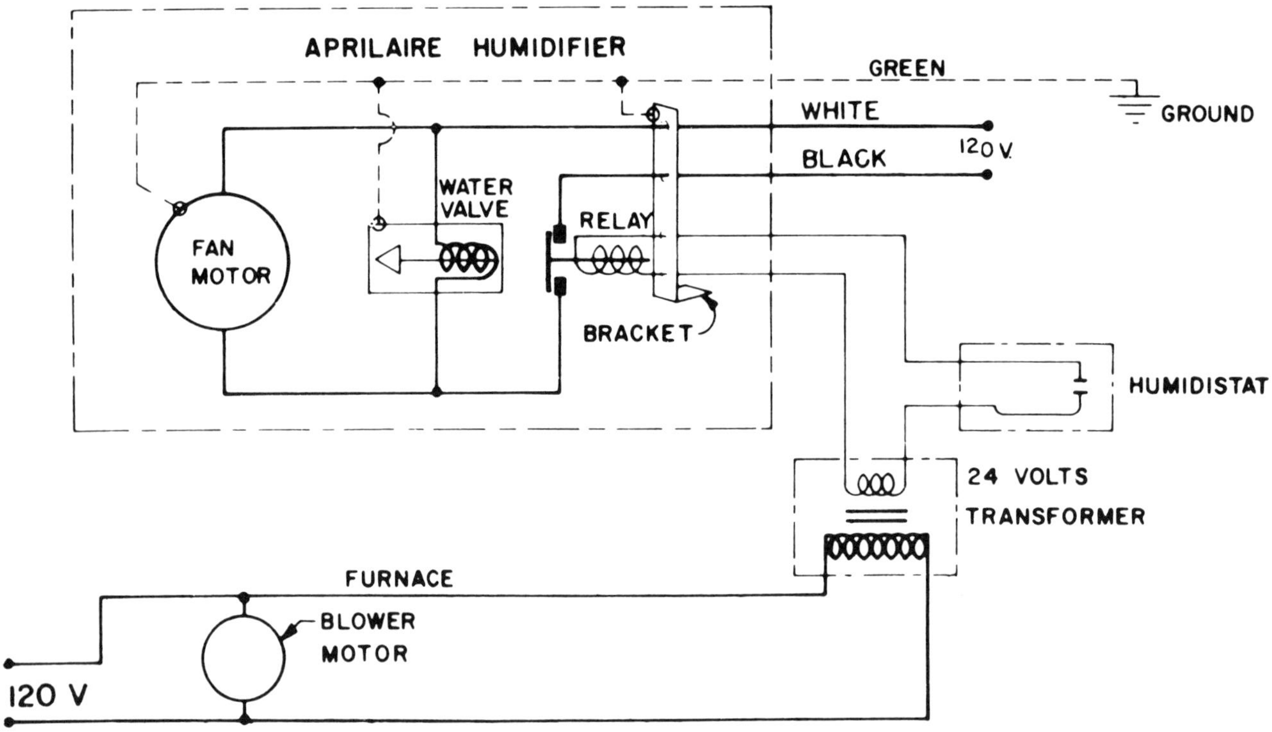

Courtesy, *Aprilaire*

With the humidifier in operation, what voltmeter readings would you expect across the following:

1. Furnace blower motor
2. Humidifier fan motor
3. Water valve solenoid
4. Electromagnetic relay coil
5. Secondary of transformer
6. Humidistat contacts
7. Relay contacts

Using the five-step procedure outlined under *Troubleshooting the Electrical Circuit*, solve these two problems:

Problem 1. Fan motor will not operate. Furnace blower motor and water valve are operating.

STEP 1. The schematic has been read.

STEPS 2 and 3.

The furnace blower motor and the water valve both operate, indicating that the fuses are good and that there is power to the humidifier.

STEP 4. The fan motor is not functioning.

STEP 5. Isolate and solve the problem using the voltmeter.

- The schematic shows that the only reasons the fan motor would not run are: either a defective fan motor or an open circuit to the fan motor.
- If the voltmeter shows 115 V across the fan motor terminals, the motor is defective
- If the voltmeter reads zero voltage across the fan motor terminals, there is an open circuit between the motor terminals and either L_1 or neutral.
- If the voltmeter indicates 115 V from one terminal of the fan motor to the neutral at the water valve connection, this line is open.
- If no voltage is indicated, read from the other terminal of the fan motor to the other connection of the water valve solenoid. You should read 115 V, indicating that this line is open.
- Remember, the water valve solenoid is operative. This means that we have voltage to this point in the circuit.

Problem 2. The water valve and fan motor will not operate. The furnace blower motor is running and the humidistat contacts are closed. Find all of the possible causes that could create this condition.

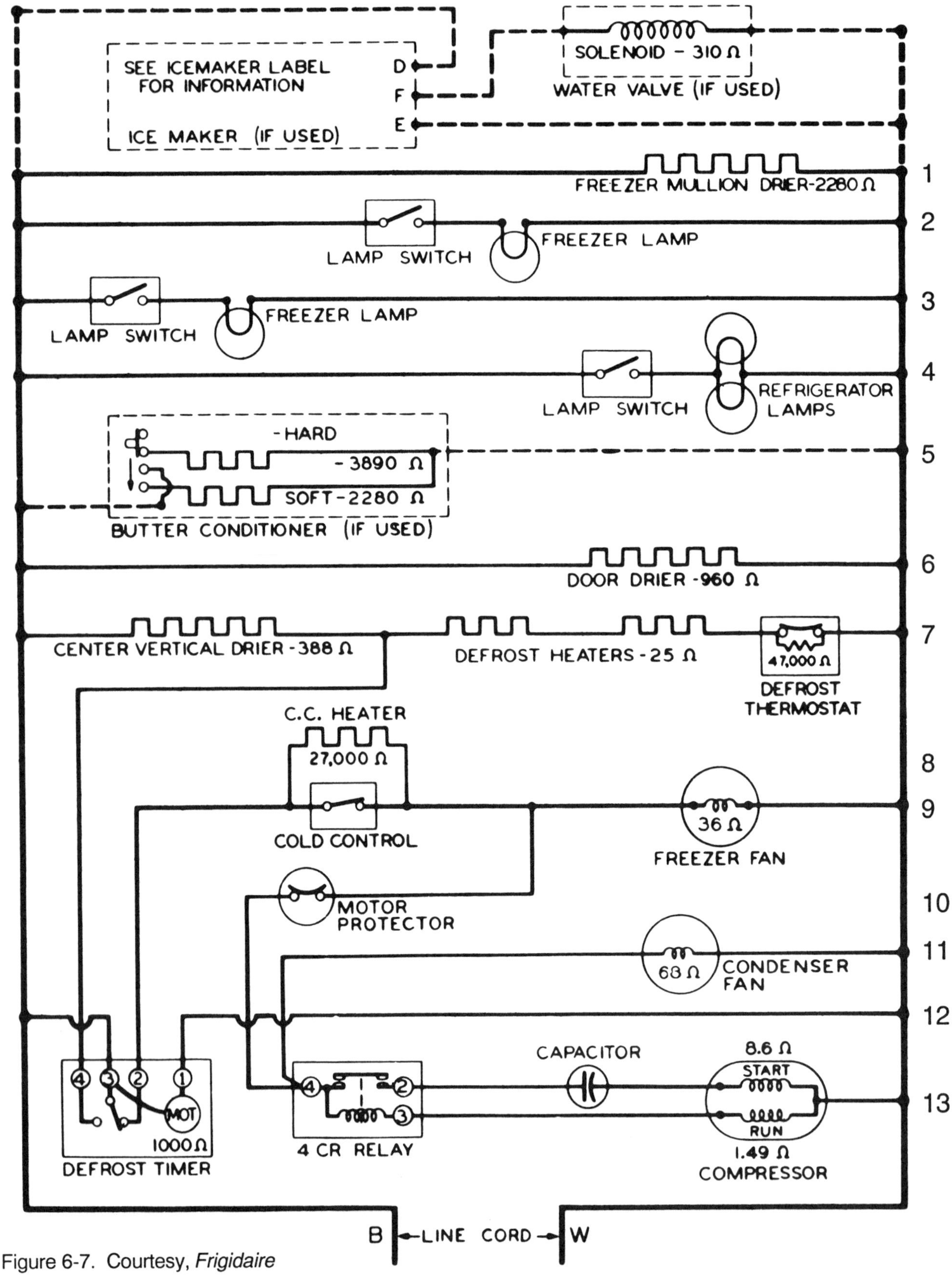

Figure 6-7. Courtesy, *Frigidaire*

FRIGIDAIRE REFRIGERATOR/FREEZER

This 115-V Frigidaire model has two separate freezer compartments and one refrigeration section with automatic defrosting. A butter conditioner and an icemaker, with its water valve, are optional with this model.

The schematic, shown in Figure 6-7, is unique in that it gives the resistance across all motors, heater coils, and the compressor windings. To the service technician, this can be a very helpful tool in troubleshooting an electrical problem.

1. GO STEP-BY-STEP THROUGH THE ELECTRICAL SEQUENCE

The easiest way to determine *what happens next* is to first eliminate the circuits that have no effect on the operation of the cooling or defrosting system.

To help in locating each device, we have numbered each of the parallel circuits down the side of the schematic. Starting from the top, the first circuit, in dotted lines, shows the field wiring connections for an automatic ice maker and its solenoid water valve.

- Line 1 shows a heater, called a freezer mullion drier  connected between B and W. (A mullion heater is used to prevent condensation forming on the strip of cabinet separating two doors. This strip is called a mullion.) The schematic shows no switches or controls. So, the mullion drier should be warm to the touch at all times.
- Line 2 shows a normally-open door switch 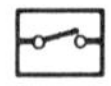 in series with the lamp 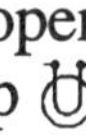 in the upper frezer compartment. The switch will close when the door is opened, completing a circuit to the freezer lamp from B to W.
- Line 3 shows another normally-open door switch controlling the lamp in the lower freezer section.
- Line 4 shows a third normally-open door switch controlling two lamps in the refrigerator section. The lamps are in series with the door switch and are wired in parallel with each other.
- Line 5 shows the field wiring for an optional butter conditioner. The butter conditioner consists of two electric resistance heaters, each with a different heat output. This added heat will hold the butter at a spreadable temperature. The slide switch has three positions:

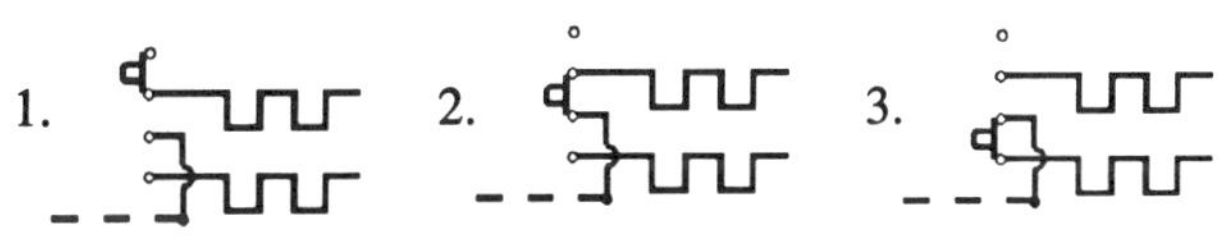

1. In the *hard* or *off* position, there is no heat added.
2. The center position completes a circuit from B, through the 3,890 ohm heater, to W.
3. In the *soft* position, a circuit is completed from B, through the 2,280 ohm heater, to W.

- Line 6 shows a resistance heater, called a door drier  which is used to keep the freezer door clear of condensation. The heater is wired from B to W, with no *on* or *off* control.

We have now disposed of all the circuits that have no effect on the cooling and defrosting cycles.

- Line 7 shows four resistors: the 388-ohm center vertical drier, a pair of defrost heaters, totaling 25 ohms, and a 47,000-ohm resistor in the defrost thermostat . Actually, the 47,000 ohm resistor has no functional purpose in this circuit. It is only there to permit continuity testing through the defrost thermostat. From now on, assume it is not in the circuit.

  The defrost thermostat senses the temperature in the refrigerator cooling coil and will close when the temperature of the cooling coil is below freezing. With the cooling coil below freezing, frost can build up on the coil and in the freezer compartments.

  When the thermostat closes, it completes the circuit to the drier and the defrost heaters. The total heat output is 32 watts, with 30 watts being concentrated at the center vertical drier. This small amount of heat prevents the formation of frost and ice at the door seals. The remaining heat, from the two defrost heaters, is of no consequence.

  The system remains in this condition until the defrost timer motor 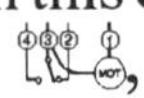, Line 13, advances to the defrost position. The motor runs constantly, as it is connected at timer terminals 1 and 3, to an uninterrupted circuit running along Line 12. The motor is programmed, by the manufacturer to start and stop the defrost cycle by closing and opening a circuit between terminals 3 and 4 in the timer.

  When terminals 3 and 4 are connected, to initiate the defrost cycle, an alternate circuit is completed from B, at Line 12, to a junction between the center vertical drier and the two defrost heaters on Line 7, thereby bypassing the center vertical drier which will not be needed while defrosting. With the 388-ohm resistor no longer in series, the two defrost heaters will now produce 530 watts. Defrosting will continue until the defrost timer interrupts the circuit or until the defrost thermostat senses that the cooling coil is above freezing and opens. This interrupts the circuit to the two defrost heaters, thereby guarding against overheating the freezer compartment during the defrost period.

  At the conclusion of the defrost period, the defrost timer opens the circuit between terminals 3 and 4 and completes the circuit between terminals 2 and 3.

- Line 8 shows a cold control heater  that is energized when the defrost timer switches to complete a circuit from B, at Line 12, through terminals 2 and 3. This heater warms the bellows of the cold control thermostat , Line 9, to ensure that the temperature is read at the sensing probe and not at the bellows.
- Line 9 shows the cold control thermostat and the freezer fan  wired in series. They operate on the same circuit as the cold control heater.

With the cold control open, the voltage drop across the cold control heater consumes virtually all the available voltage. Consequently, the freezer fan will not operate.

When the cold control is closed, calling for refrigeration, the cold control heater is bypassed and the freezer fan, mounted near the cooling coil, will circulate cold air through all compartments of the refrigerator.

With the cold control switch closed, a parallel circuit is completed through the motor protector, Line 10, to terminal 4 of the compressor relay (CR), Line 13. (On some current relays, terminal 4 is marked L.)

Simultaneously, another parallel circuit is completed from terminal 4 of the CR relay to the condenser fan on Line 11.

- Line 13 shows a circuit completed from terminal 4 in the CR relay, through the relay coil, to terminal 3 and through the compressor run winding to W. With the relay coil energized, the contacts close, completing a circuit through terminal 2, to the capacitor and the compressor start winding. (On some current relays, terminals 2 and 3 will be marked S and M.) When the compressor is nearly up to speed, the relay will open and take the start winding out of the system.

The compressor will continue to run as long as the cold control remains closed, calling for refrigeration, or until the defrost timer advances to the defrost cycle, thereby opening terminals 2 to 3 in the timer and closing terminals 3 to 4.

During the defrost cycle, the center vertical drier will again be bypassed and the defrost thermostat will initiate defrosting when it senses temperatures below 32 °F.

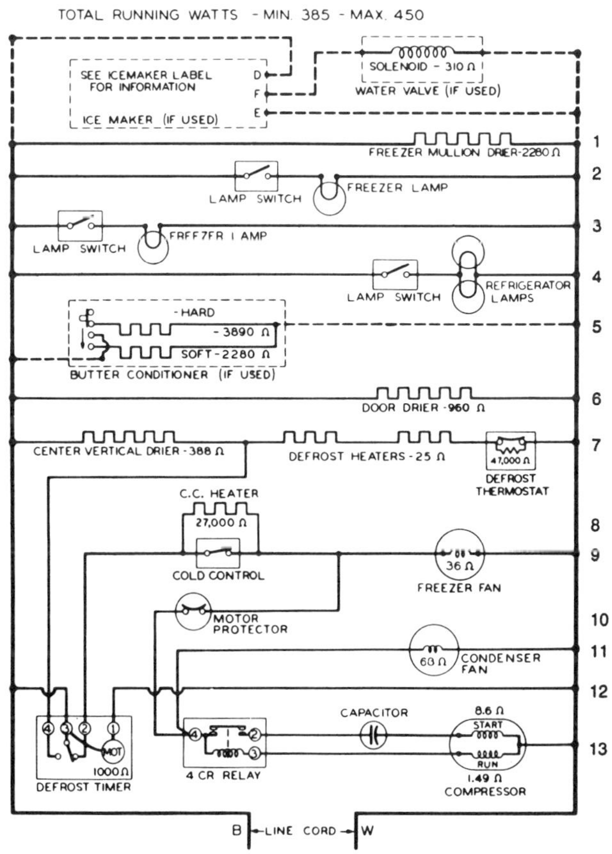

Courtesy, *Frigidaire*

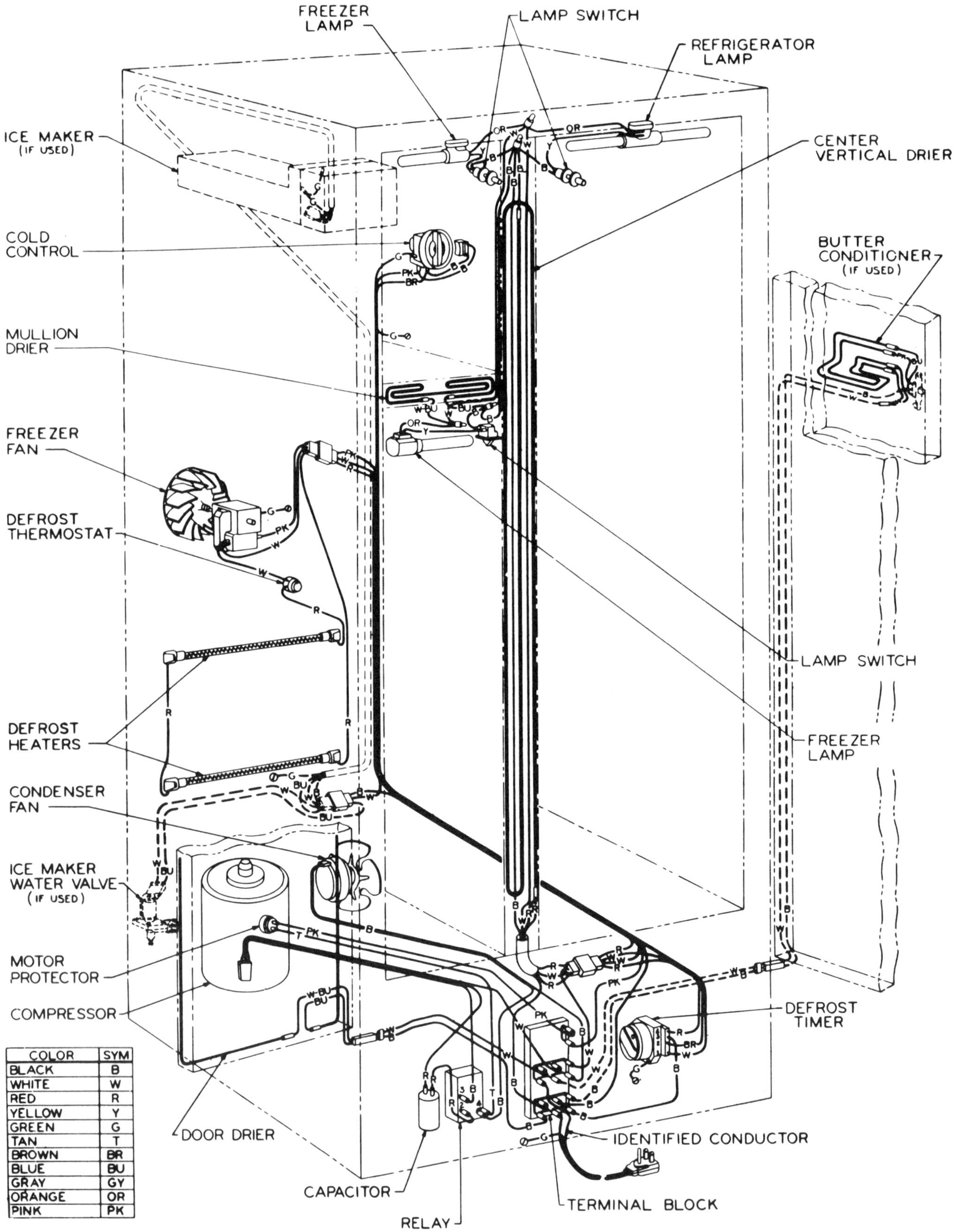

COLOR	SYM
BLACK	B
WHITE	W
RED	R
YELLOW	Y
GREEN	G
TAN	T
BROWN	BR
BLUE	BU
GRAY	GY
ORANGE	OR
PINK	PK

Courtesy, *Frigidaire*

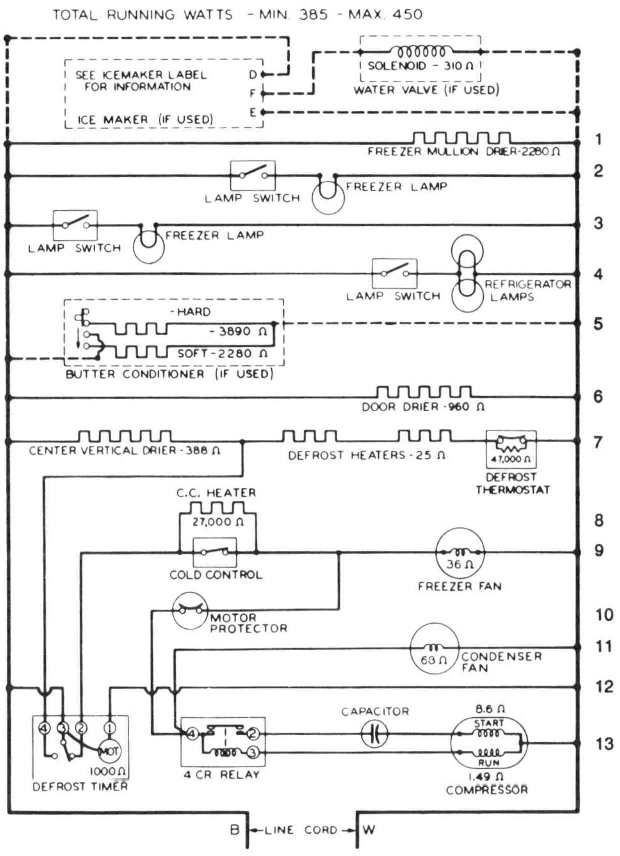

Courtesy, *Frigidaire*

What voltages would you expect to read across the following motors, heaters and switches; first in the cooling cycle and then in the defrost cycle:

1. Freezer mullion drier, Line 1
2. Door drier, Line 6
3. Center vertical drier, Line 7
4. Defrost heaters, Line 7
5. Cold control heater, Line 8
6. Freezer fan, Line 9
7. Condenser fan, Line 11
8. Defrost timer motor, terminals 1 to 3, Line 13
9. Common to run on the compressor, Line 13
10. Cold control, Line 9
11. Motor protector, Line 10

Solve these problems using the five-step procedure.

Problem 1. Refrigerator will not defrost.

Problem 2. Refrigerator will not come out of defrost.

Problem 3. Compressor will not start and run. There is no hum at the compressor. Condenser fan and freezer fan are working.

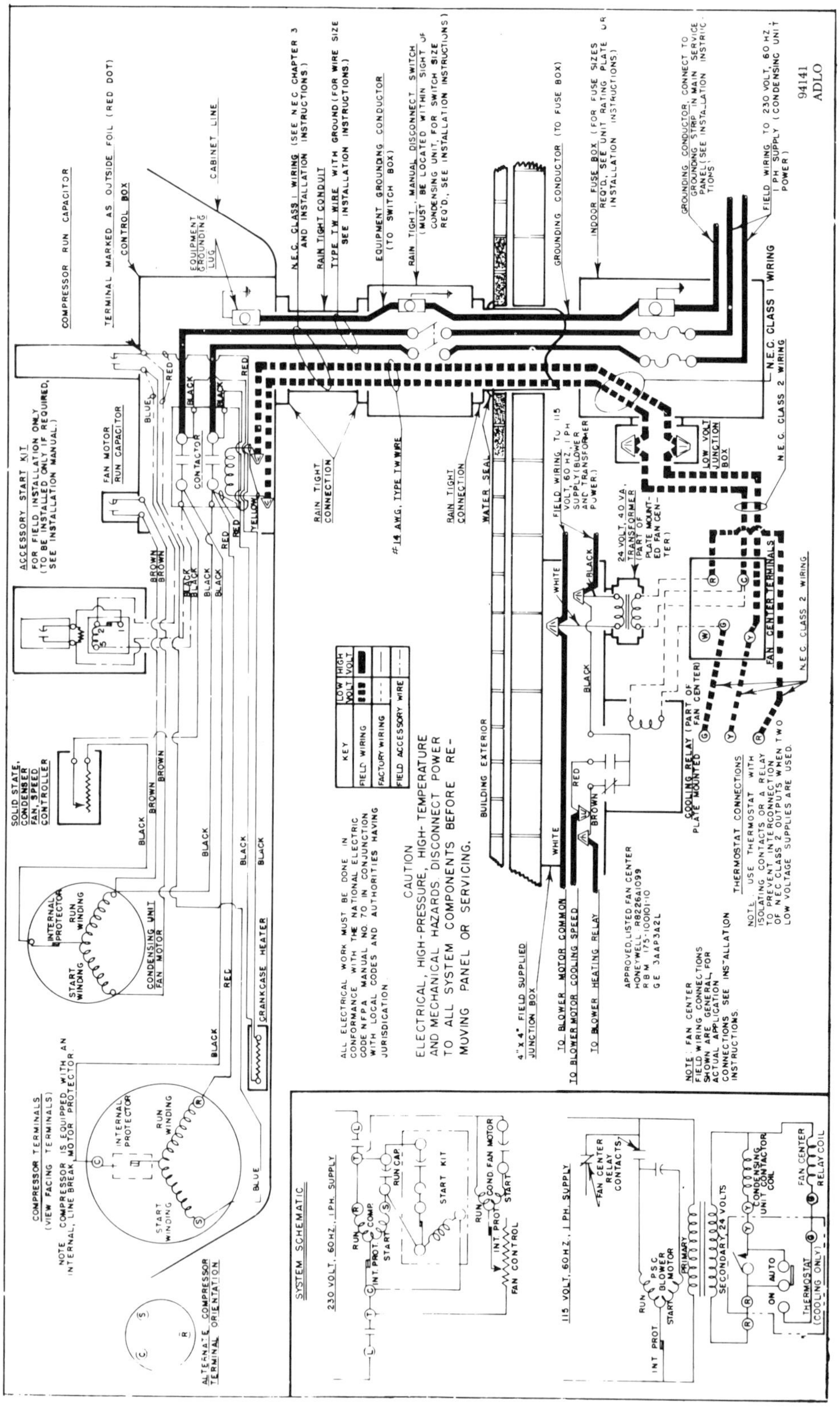

Figure 6-8. Courtesy, *Heil-Quaker*

HEIL-QUAKER AIR CONDITIONING CONDENSING UNIT

Reading from top to bottom, the schematic for the Heil-Quaker condensing unit, Figures 6-8 and 6-9, shows the high-voltage (230 V) outside wiring, 115-V furnace wiring, and the 24-V control wiring. The condensing unit contactor coil, shown on the 24-V inside wiring schematic, is actually located in the outside condensing unit. Field wiring — — —, from the thermostat, is run to the contactor coil.

With an add-on system, an A coil, or cooling coil, is installed in the plenum of a hot air furnace. The wiring to the outside condensing unit includes both a 230-V and a 24-V circuit.

For this example, we will only cover the condensing unit.

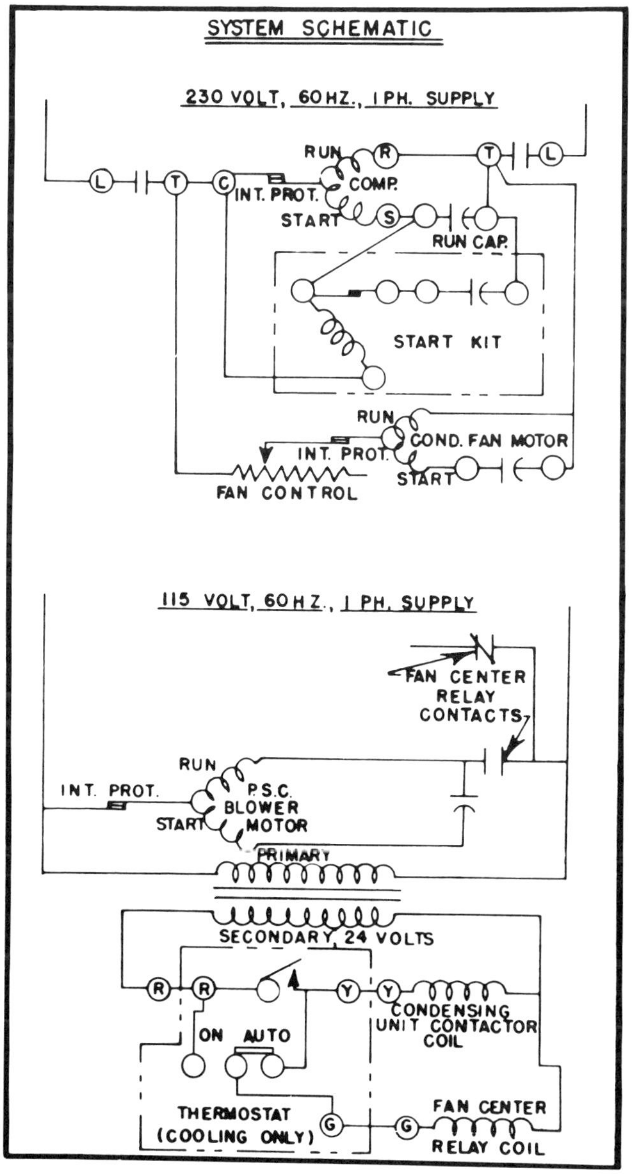

Figure 6-9. Courtesy, *Heil-Quaker*

1. Go Step-by-Step Through the Electrical Sequence

To determine *what happens next*, set the thermostat selector to COOL, the fan switch (ON | AUTO) to AUTO, and lower the thermostat setting below room temperature to close contacts (R) to (Y).

- With contacts R to Y closed, a circuit is completed from the 24-V transformer secondary (SECONDARY 24 VOLTS) and through the condensing unit contactor coil (CONDENSING UNIT CONTACTOR).
- Since the fan control contacts (ON | AUTO) have been manually closed, the indoor fan relay coil (FAN CENTER) is energized whenever the thermostat contacts close for cooling. This ensures that the indoor fan will run whenever the compressor starts. With the indoor fan relay coil energized, the normally-open fan center relay contacts (FAN CENTER) close, completing a circuit through the internal protector (INT. PROT.), to the permanent split capacitor (PSC) blower motor.
- When the condensing unit contactor coil (CONDENSING UNIT CONTACTOR) is energized, the normally-open contacts (L)-||-(T) and (T)-||-(L), in the 230-V circuit close, completing two parallel circuits.
- One circuit, from T to T on the contactor, goes through the solid state condenser fan speed control (FAN CONTROL). The speed control matches the speed of the condenser fan motor to the outside air temperature. As the outside temperature increases, the condenser fan runs faster. A decrease in outside air temperature slows the fan motor. The speed control automatically varies the fan speed when the outdoor air temperature is between 75 °F and 105 °F.

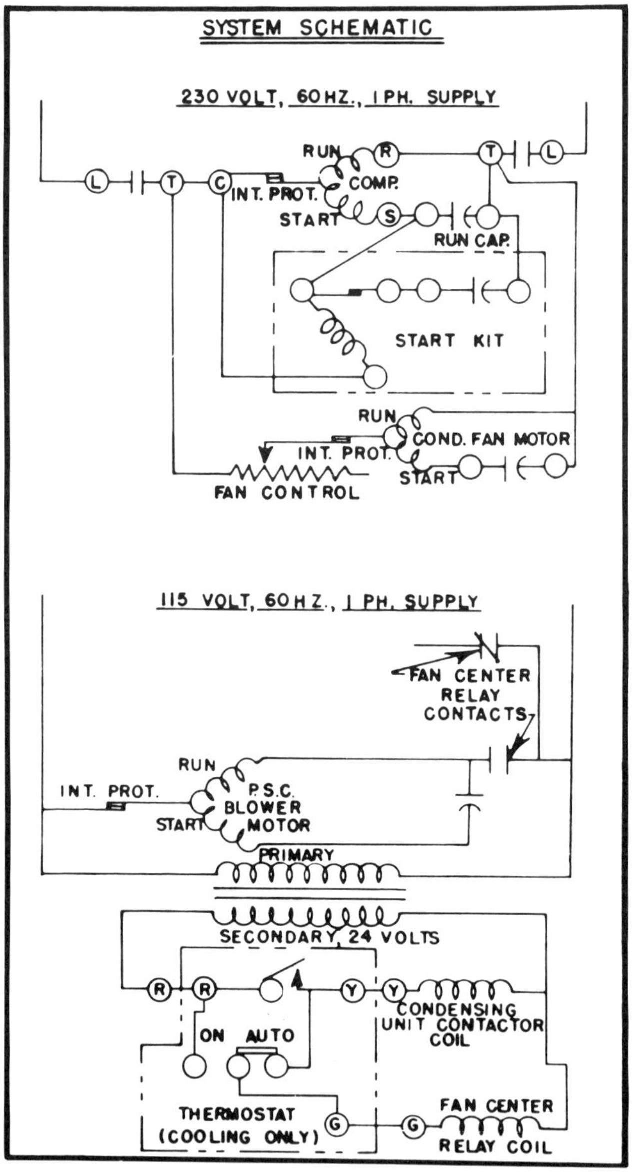

Courtesy, *Heil-Quaker*

- From the speed control, the circuit continues through the internal protector to the common of the condenser fan motor. Even though the internal protector is schematically shown outside the fan motor, it is located inside the motor and connected between the outside terminal of the motor and the common junction of the start and run windings.
- With the contactor contacts closed, a second circuit is completed through the internal protector to the common in the compressor.
- This compressor has a permanent split capacitor motor (PSC) with a run capacitor. The schematic shows where to wire a hard-start kit, if required.

The hard-start kit, shown inside the dotted line, consists of a potential start relay and a start capacitor. The kit is used to overcome hard starting in areas where voltages are below normal.

Observing the condenser schematic:

- The condensing unit contactor controls the operation of both the compressor and the condenser fan motor.
- The condenser fan motor can operate without the compressor and the compressor can operate without the fan motor. This can only happen if one of the motors is defective.
- So long as there is 24 V to the condensing unit contactor coil, the condensing unit fan and compressor can operate.

Set the thermostat to start the system. Using a voltmeter, what voltages would you expect to read across the following motors, contacts and coils:

1. Ⓡ to Ⓒ on the compressor
2. RUN to common on the condenser fan motor
3. Ⓛ to Ⓣ on the contactor
4. Ⓡ to Ⓨ on the thermostat
5. Ⓒ to Ⓨ on the condensing unit con tactor coil
6. Primary of transformer

Using the five-step procedure, solve these three problems:

Problem 1. Condenser fan does not run. Compressor is running.

Problem 2. Condenser fan and compressor are not running. Blower motor is operating.

Problem 3. Condenser fan, compressor and blower motor will not run.

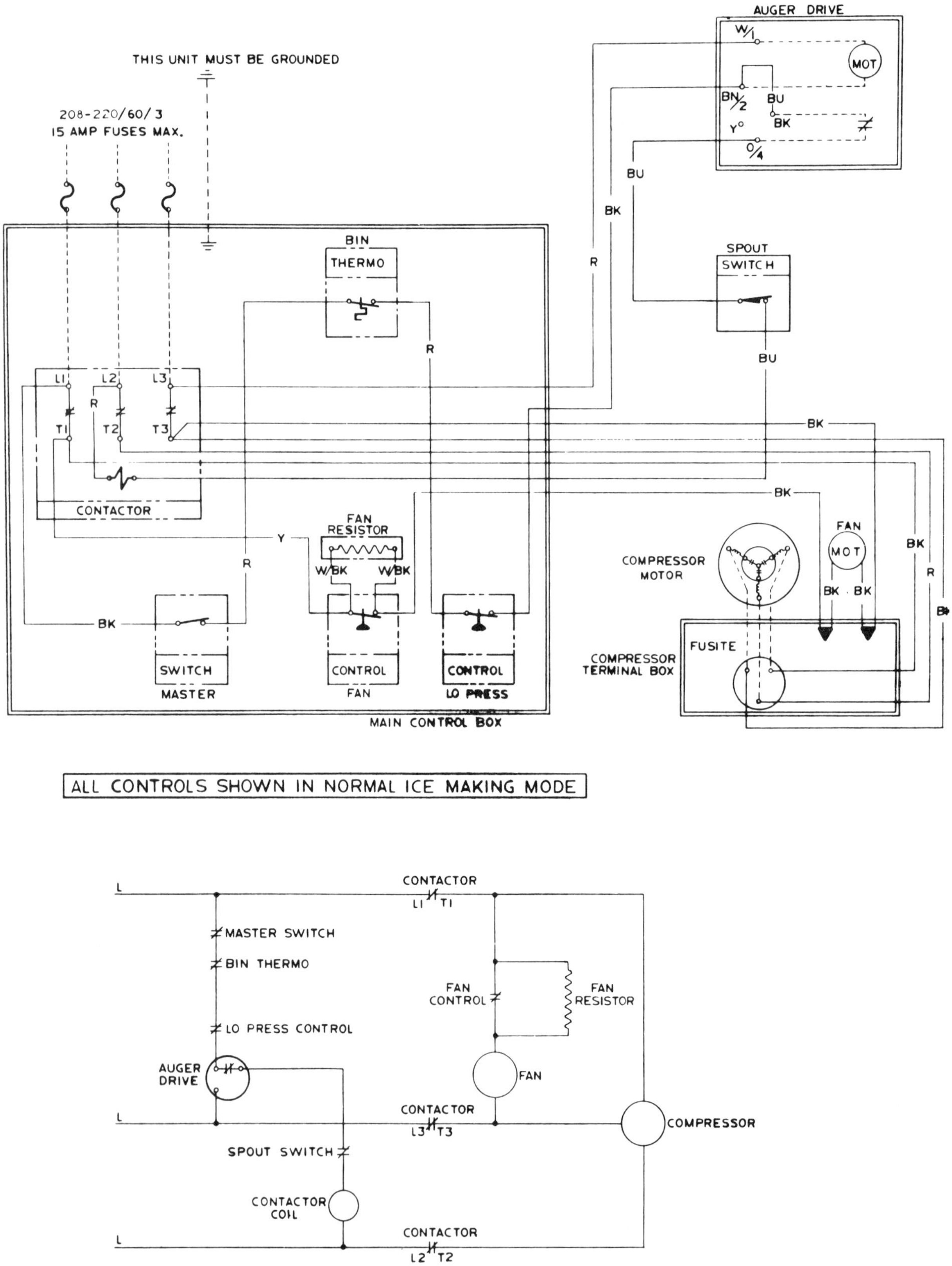

Figure 6-10. Courtesy, *Scotsman*

SCOTSMAN COMMERCIAL ICE FLAKER

This Scotsman Ice Flaker, Figure 6-10, delivers a high volume and continuous flow of flaked ice to the storage bin.

Like most ice makers, the flaker will continue to make ice until the bin is full or the master switch is opened. The bin thermostat sensing element, mounted near the top of the bin, is set to open the thermostat contacts at approximately 32 °F, the temperature of ice, and shut down the system. If ice is constantly removed from the bin, the flaker will continue to make ice.

1. GO STEP-BY-STEP THROUGH THE ELECTRICAL SEQUENCE

The wiring diagram shows the compressor to be 208 / 220 V, 3ϕ, 60 Hz. The condenser fan motor, auger motor and compressor contactor coil are 208 / 220 V, 1ϕ, 60 Hz. Since the schematic in Figure 6-16 is rudimentary, showing all switches as , we will amplify this information by showing the graphics from the wiring diagram while following the schematic.

All contacts are in the operating or ice-making position, even though the normal position of some contacts, such as the low pressure switch , contactor contacts , fan control , and master switch , is open.

- To start the ice machine, the manually-operated master switch must be closed, initiating a circuit from L1 through the master switch to the bin thermostat .
- The bin thermostat, which opens on a temperature fall, would be closed at this time, permitting the circuit to continue to the low-pressure control .
- The low-pressure control serves two purposes. First, it shuts down the system if the temperature and pressure of the refrigerant in the evaporator drops too low. This drop could be due to a loss of water, which is the heat load on the evaporator. A drop in evaporator pressure could cause the turning auger to freeze, with possible damage to the auger drive and the auger itself.

  Second, because the compressor will continue to operate until the bin thermostat opens, a loss of refrigerant will keep the compressor running even though no ice is being produced. The low-pressure switch will open if there is a loss of refrigerant.
- From the low-pressure control, the control circuit runs through a normally-closed motor protector switch , that protects the auger motor.

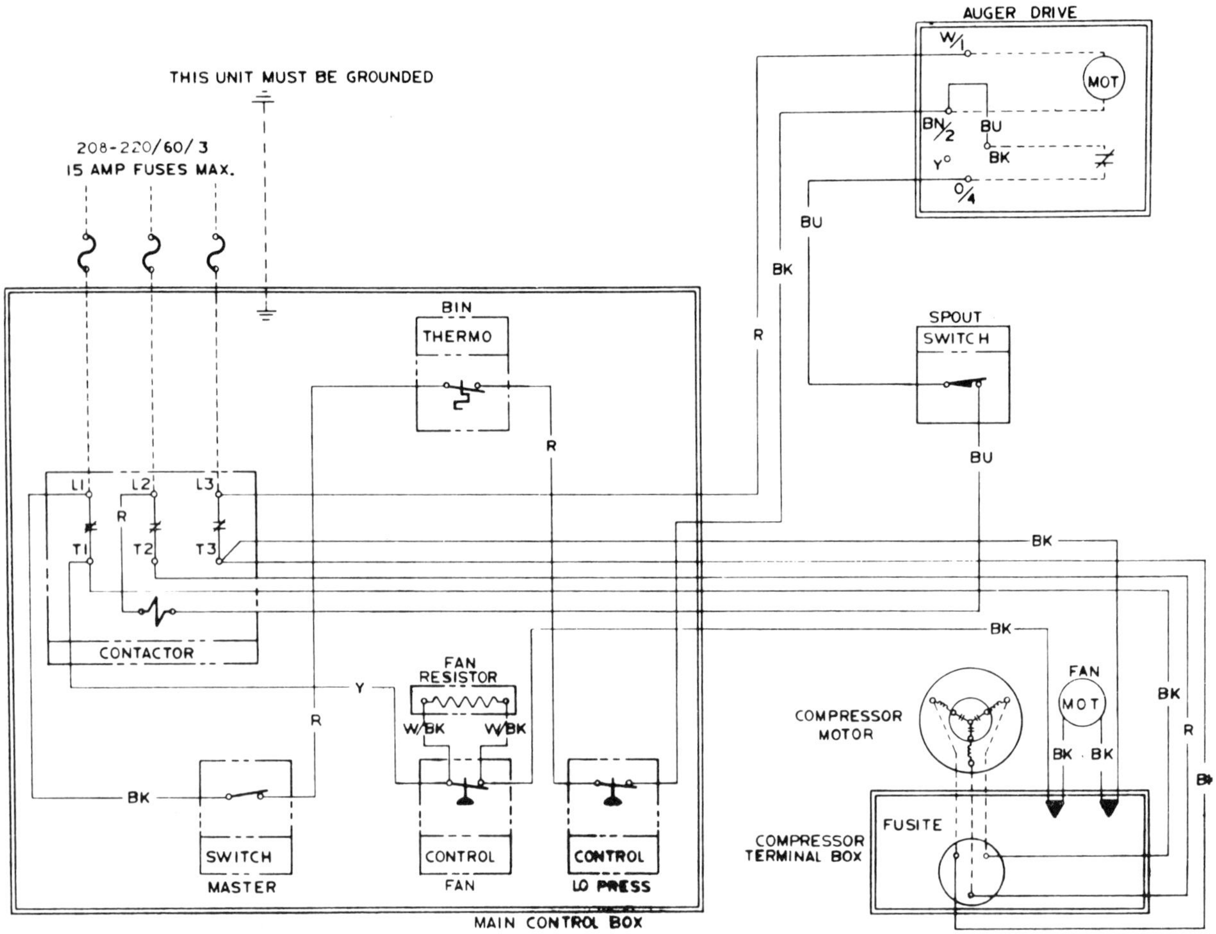

ALL CONTROLS SHOWN IN NORMAL ICE MAKING MODE

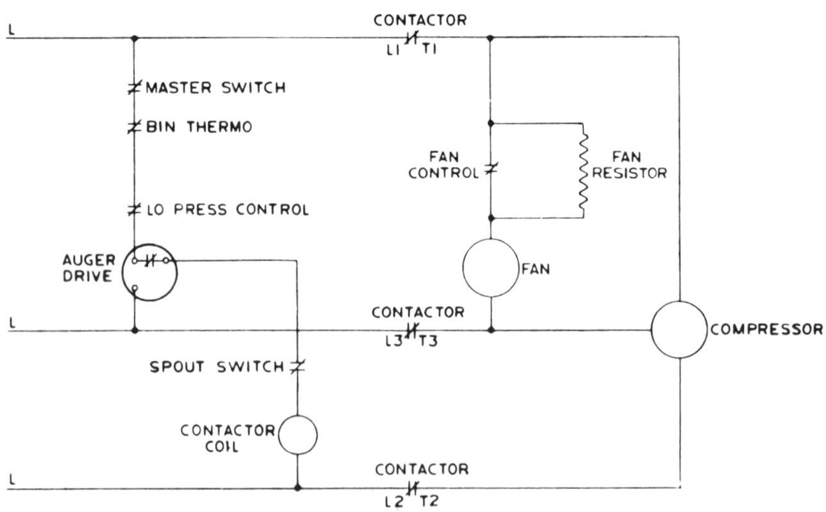

Courtesy, *Scotsman*

- The circuit continues through the normally-closed spout switch. The spout switch is a mechanically-operated safety that will open if ice builds up in the spout, pushing the switch open.
- From the spout switch, the circuit energizes the compressor contactor coil and terminates at L2.
- The auger drive motor, wired from L1 to L3, is controlled by the master switch bin thermostat, and low-pressure control and will continue to run if the spout switch is open, or if an open contactor coil failed to close the contactor contacts.
- With the compressor contactor contacts closed, L1 T1, L2 T2, L3 T3, the three-phase compressor will start. However, the compressor will not start if either the master switch, bin thermostat, low-pressure control, auger overload or spout switch are open. The control circuit is a series circuit.
- The condenser fan motor, wired between F1 and T3, will run when the compressor contactor is closed. This is to guarantee that the condenser fan and compressor operate together.
- Between T1 and the condenser fan motor is a fan control, and a fan resistor that is wired parallel to the fan control. The fan control is a low-pressure switch that makes on an increase in head pressure. If the air temperature at the condenser is low, the fan control will be open and the fan will be energized through the fan resistor, causing the fan to run at a low speed. When the air temperature increases, the head pressure will rise, closing the fan control. The condenser fan will then run at high speed.

Observing the schematic: If fuse L2 is out, the auger motor will still operate. If fuse L3 is out, the contactor will pull in, but the compressor, condenser fan, and auger will not start. If fuse L1 is out, nothing will start.

What voltages would you expect to read across the following switches, motors and coils, with the system operating on a primary voltage of 220 V, 3ϕ:

1. Master switch
2. Bin thermostat
3. Low-pressure control
4. Spout switch
5. Fan control (motor at low speed)
6. Fan motor (motor at high speed)
7. Auger motor
8. Contactor coil

Using the five-step procedure, solve the following problems.

Problem 1. Fan motor will not operate. Compressor and auger motor running.

Problem 2. Compressor and fan not starting. Auger motor running.

Problem 3. Nothing operating. 220 V at L1, L2, L3.

ARKLA AIR-COOLED CHILLER

The Arkla air-cooled chiller, Figure 6-11, is an ammonia/water absorption refrigeration system, using natural or LP gas as the heat source for the generator. Electricity performs a supportive role by operating the controls, gas valve, chilled water pump and the solution pump.

The absorption chiller removes heat from recirculated water. The chilled water, at about 45 °F, is then circulated through a remote blower coil. Return air is drawn across the coil, giving up heat to the chilled water. The cooler air is discharged into the conditioned space and the warm water, at about 55 °F, is returned to the chiller.

The wiring diagram (not shown) appears to be loaded with controls. The schematic in Figure 6-17, however, shows only a few necessary electrical devices. The chiller comes completely factory wired, except for the low-voltage thermostat. Primary voltage, supplied to L1 and L2, is 230 V, 1ϕ, 50 or 60 Hz.

1. GO STEP-BY-STEP THROUGH THE ELECTRICAL SEQUENCE

Set the thermostat selector to COOL and fan on AUTO. (The thermostat and the indoor fan circuit are not shown on this schematic. The fan will start when the thermostat contacts close for cooling.) Lower the thermostat temperature setting below room temperature to complete the circuit from (R) to (Y).

- With R to Y closed, a circuit is completed from the transformer secondary to the heater in the time-delay switch and from there, to the high-temperature switch. The high-temperature switch is a limit switch that prevents overheating.
- After approximately 45 seconds, the time-delay heater will close the time-delay switch and energize the relay coil.
- With the relay coil energized, the relay contacts in the 230-V circuit will close and start both the chilled water pump motor (split phase, capacitor start) and the condenser fan motor (permanent split capacitor).
- With the fan operating, air movement will close the normally-open sail switch and complete a parallel circuit, through the normally-closed chilled water switch to the gas valve in the 24-V system.

The purpose of the sail switch is to prevent overheating by ensuring that the condenser fan is providing vent and draft for the burner.

The purpose of the chilled water switch is to prevent freeze-up if the water is not being circulated. This switch opens at 38 °F, stopping the absorption cycle, and resets at 42 °F.

- With the circuit completed to the gas valve, the burner can be ignited and the absorption refrigeration cycle begins.
- If the thermostat opens, or if one of the safety switches opens, there will be no voltage to either the gas valve circuit or the time-delay switch heater. However, residual heat from the heater will hold the time-delay switch closed for approximately 3-1/2 minutes during which the pump and fan will continue to operate. When the time-delay switch has cooled, the switch will open and all circuits will be de-energized.

Set the thermostat selector to COOL, fan to AUTO and lower temperature setting to close R to Y. After 45 seconds, what voltages would you expect to read across the following motors, coils and switches:

1. Relay contacts
2. Time-delay switch
3. Sail switch
4. Chilled water switch
5. High-temperature switch
6. Gas valve coil
7. Time-delay relay coil
8. Relay coil
9. Pump motor
10. Run to common on fan motor

Using the five-step procedure, solve these problems:

Problem 1. Gas valve will not open, pilot is on.

Problem 2.. Fan and pump will not start.

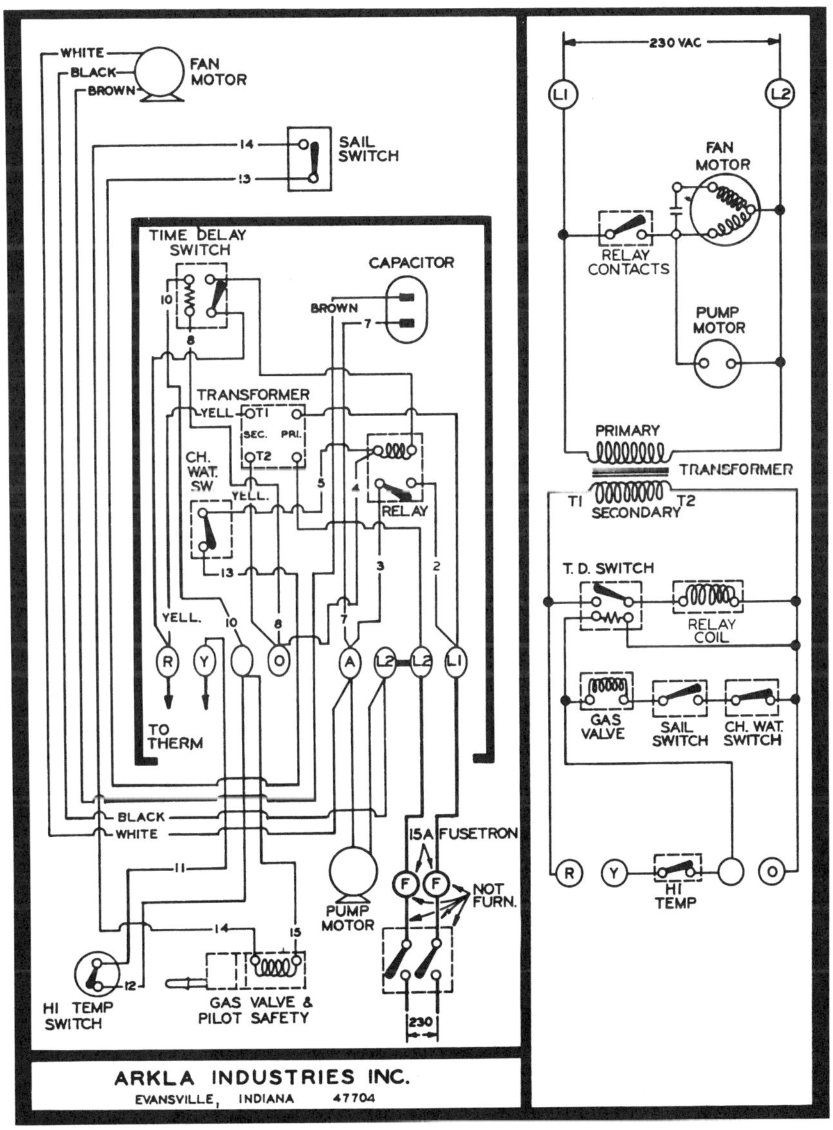

Figure 6-11. Courtesy, *Arkla*

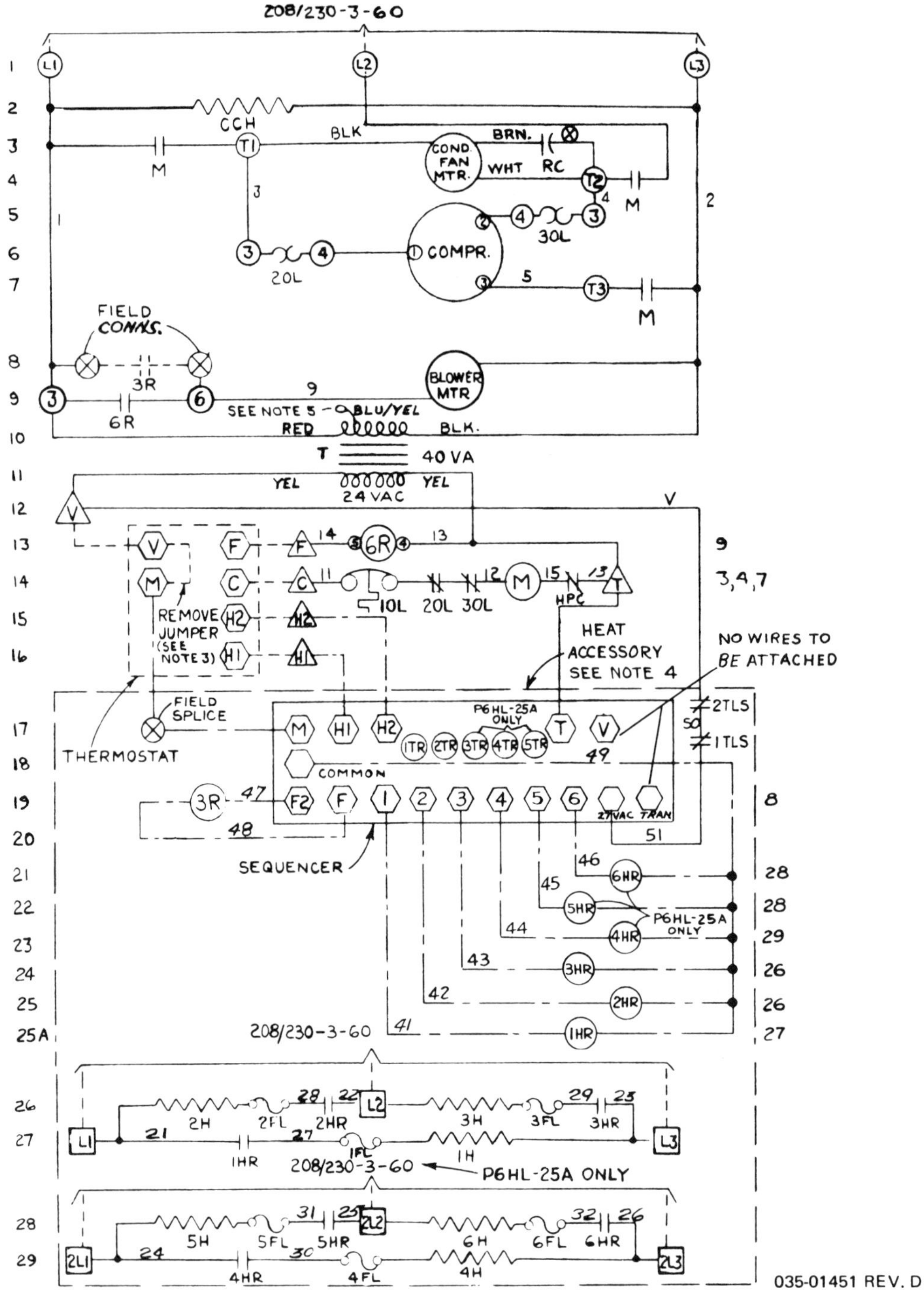

LEGEND

M	COMPRESSOR CONTACTOR	TR	TIME DELAY RELAY*
3R	FAN RELAY, HEAT*	HPC	HIGH PRESSURE CONTROL
6R	FAN RELAY, COOL	OL	OVERLOADS
HR	HEAT RELAY*	△	LOW VOLTAGE TERMINAL
H	HEATER ELEMENT*	□	POWER WIRING TERMINAL
FL	FUSIBLE LINK*	○	ACCESSORY TERMINAL
RC	RUN CAPACITOR	⊗	FIELD SPLICE OR CONNECTION
TLS	TEMPERATURE LIMIT–SWITCH*	———	FACTORY WIRING
CCH	COMPRESSOR CRANKCASE HEATER	— — —	20 V DC WIRING
T	L.V. TRANSFORMER W/FUSED PRIMARY	- - - -	FIELD WIRING

*PART OF HEAT ACCESSORIES

Figure 6-12. Courtesy, *York*

YORK PACKAGE AIR CONDITIONER WITH ELECTRIC RESISTANCE HEAT

The York schematic, Figure 6-12, is a little different from the previous examples. Note that the numbers running down the left side of the schematic are for line reference only. The lines on the right side, however, indicate the lines on which the contacts for each relay are found.

For example, reading from the left across Line 14, you will find a relay coil (M). Continuing across Line 14, to the right margin, you will find the notation 3, 4, 5, indicating that the contacts for relay M are found on Lines 3, 4 and 7.

The room thermostat for this unit should have separate heating and cooling circuits. The sequencer, used to bring in the heating coils, converts 24 vac to 20 vdc and is equipped with heater relays, time-delay relays and a fan relay, all operating on 20 vdc. The room thermostat jumper, between terminals ⟨V⟩ and ⟨M⟩ (Lines 13 and 14), must be removed. If not removed, it will short out ac to dc power and damage the sequencer. The thermostat may be either single-stage heat and single stage cool or two-stage heat and single-stage cool.

Referring to the schematic,. Lines 14, 15 and 16, the thermostat terminals ⟨M⟩, ⟨H2⟩ and ⟨H1⟩ are dc and connected to the sequencer at Line 17. All wires numbered 41 through 51 are direct current. All other wires are alternating current.

The number of each wire is actually printed on the wire insulation. For instance: relay (3HR), found on Line 24, has its contacts on Line 26 and is connected to ⟨3⟩ on Line 19 by wire number 43.

The operation of the solid state heating sequencer will be discussed when we trace the heating cycle. First, we will trace the cooling system.

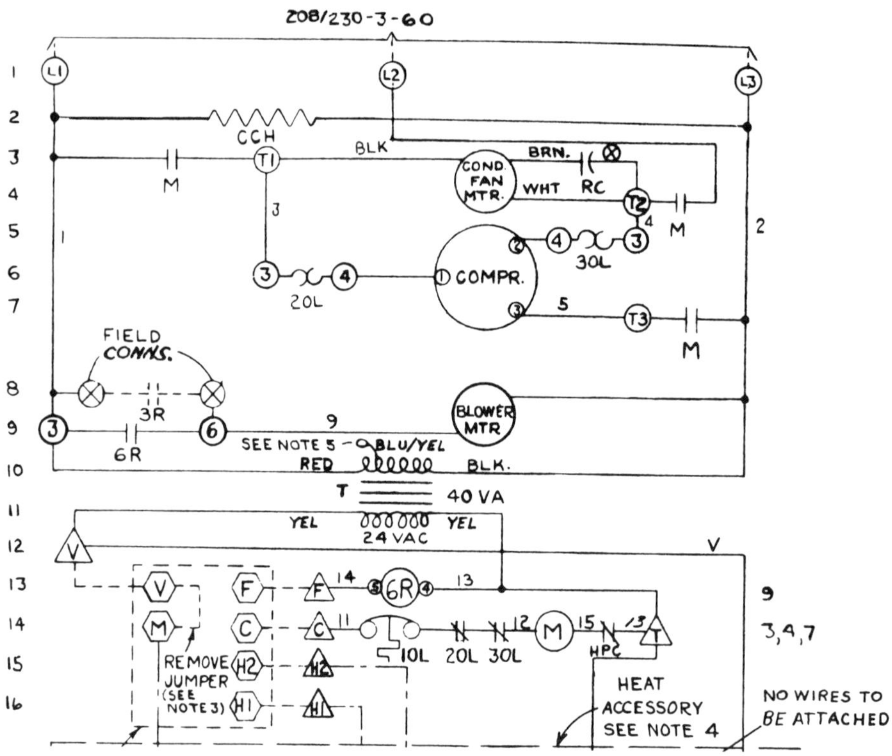

Courtesy, *York*

3. Go Step-by-Step Through the Electrical Sequence

COOLING CYCLE

- Place thermostat selector on COOL, fan on AUTO. Lower the thermostat setting below room temperature to close contacts ⟨V⟩ to ⟨C⟩, Line 13 to Line 14.
- With contacts ⟨V⟩ to ⟨C⟩ closed, a circuit is completed from /V\, Line 12, to one side of the 24-V transformer secondary (24 VAC), Line 11, to /C\, Line 14, the low-voltage terminal in the package unit.
- From /C\, the circuit continues, on Line 14, through the temperature-actuated overload 10L, and the current-actuated overloads 20L and 30L, energizing the compressor contactor coil (M).
- The circuit from the compressor contactor coil is completed through the normally-closed high-pressure control HPC, to the low-voltage terminal /T\, the other side of the 24-V transformer.
- When thermostat contacts ⟨V⟩—⟨C⟩ close, the circuit through contacts ⟨V⟩—⟨F⟩, Line 13 is energized, completing a circuit to terminal in the package unit.
- From /F\, the circuit continues to the cooling fan relay coil (6R) and back to the 24-V transformer secondary.

Observing the low-voltage cooling control circuit:

- The compressor contactor coil (M) will de-energize if the overloads 10L, 20L, 30L or the high-pressure switch HPC, all on Line 14, open or if the contacts V to C in the thermostat open.
- The cooling fan relay coil will de-energize only by opening contacts V to F in the thermostat.

Recheck the low-voltage circuit before tracing the 3-phase high-voltage circuit:

- Wired between L1 and L3 on Line 2 is a compressor crankcase heater CCH. Since there are no switches in the crankcase heater circuit, the heater is hot as long as there is 230 V at L1 and L3.
- When compressor contactor coil (M) is energized, the margin note to the right of Line 14 indicates that the contacts M on Lines 3, 4, and 7 are closed.

  Line 3 shows normally-open compressor contactor contacts M. With these contacts closed, a circuit is completed from L1-T1 to the common terminal of the condenser fan motor (COND. FAN MTR). The schematic indicates that the condenser fan motor is a permanent split capacitor (PSC), with a run capacitor RC. A circuit is also run from L1-T1, through overload protector 20L to compressor terminal 1.
- Line 4 shows normally-open contacts M. When these contacts close, a circuit is run from L2-T2, to the run and start terminals of the condenser fan motor. A circuit is also completed from L2-T2, through an overload protector 30L to compressor terminal 2.
- Line 7 shows normally-open compressor contactor contacts M. With these contacts closed, a circuit is completed from L3-T3 to terminal 3 on the compressor.
- The contacts 6R on Line 9 close when the cooling fan relay coil (6R) is energized, completing a circuit to the blower motor (BLOWER MTR) from L1 to L3.
- The 208/230-V step-down control circuit transformer BLU / YEL, Line 10, is wired hot from L1 to L3.

Observing the high-voltage cooling control circuit:

- The condenser fan motor, wired between L1 and L2 will not run if fuse L3 is open. The fuses, which precede L1, L2, and L3, are not shown. If L3 is open, the transformer, supplying voltage to coil (M), is de-energized.
- The blower motor, wired between L1 and L3, will run if fuse L2 is open.
- If L1, L2 or L3 is open, the compressor will not run.
- The control transformer and the crankcase heater, wired between L1 and L3, will operate if fuse L2 is open.
- In a 3-phase circuit, it is necessary to check each lead, L1, L2, and L3, with a voltmeter, before attacking a problem. It is easy to be fooled into thinking that the nonoperating motor or device is defective where, in reality, one of the fuses is open.

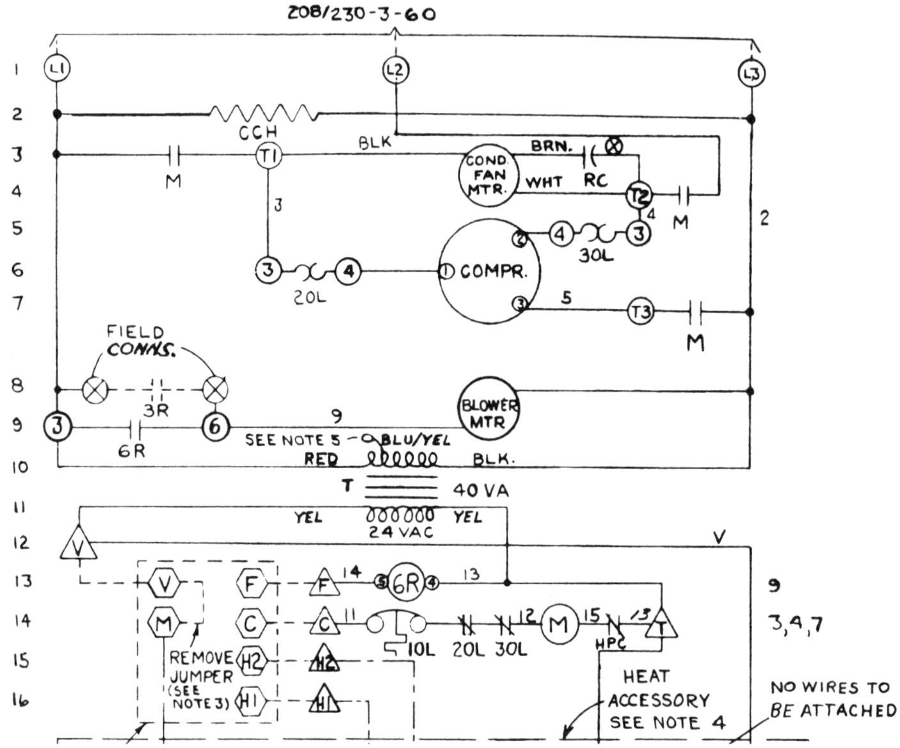

Courtesy, *York*

With the thermostat set on COOL, the fan control on AUTO (to close contacts V-F), and the temperature setting lowered (to close contacts V-C), what voltages would you read across the following cooling side components:

1. Coil CCH (Line 2)
2. Coil (M) (Line 14)
3. Coil (6R) (Line 13)
4. Contacts M (Lines 3, 4, 7)
5. Contacts HPC (Line 14)
6. Contacts 1OL, 2OL, 3OL (Line 14)
7. Contacts 6R (Line 9)
8. Compressor Terminals 1 to 2, 2 to 3, 1 to 3
9. Condenser Fan Motor Terminals T1 to T2
10. Blower Motor Terminals

Solve these problems using the five-step procedure:

Problem 1. Blower motor will not run. Condenser fan motor and compressor are operating.

Problem 2. Condenser fan and compressor not running, blower motor operating.

Problem 3. Compressor not running. Condenser fan and blower motor running.

HEATING CYCLE

The electric heat accessory package for this unit is a solid-state sequencer. This sequencer has several advantages, such as quiet operation, time delay between each heat strip, continued fan operation until all heater strips have been de-energized, and many staging possibilities.

The sequencer converts 24 vac to 20 vdc. All relays contained in the sequencer are controlled by 20 vdc. It is essential that the proper thermostat be used, which has separate heat and cool circuits, to ensure circuit isolation.

- ⟨M⟩, ⟨H2⟩, and ⟨H1⟩ on the thermostat are connected to terminals ⟨M⟩, △H2, and △H1 on the sequencer using 20 vdc field wiring.
- △V on one side of the 24 vac transformer, is factory wired to terminal ⬡ 27 VAC on the sequencer. Another wire from △T, the other side of the transformer, is factory wired to terminal ⟨T⟩ on the sequencer. From terminals ⬡ 27 VAC and ⟨T⟩, the sequencer converts the 24 vac to 20 vdc. The 20 vdc circuit is isolated from the 24 vac circuit by removing the jumper wire between ⟨V⟩ and ⟨M⟩ in the thermostat. It is necessary to use a dc voltmeter for checking the heating control circuit, and an ac voltmeter for checking the cooling circuit.

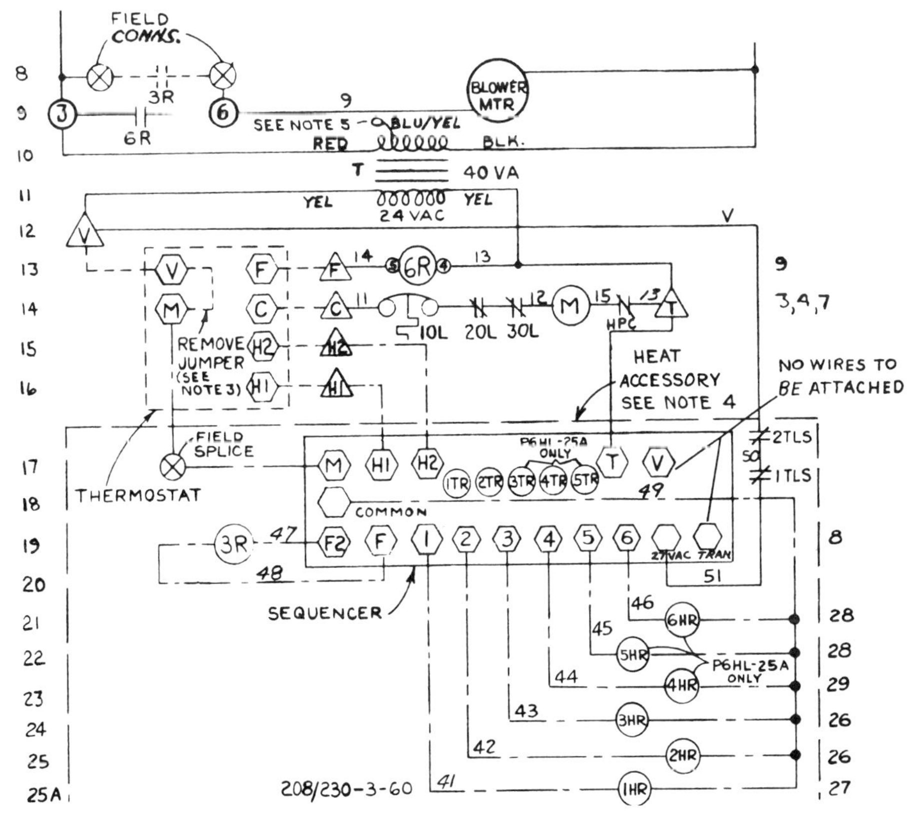

Courtesy, *York*

Set the thermostat selector on HEAT, fan on AUTO, and raise the temperature setting to close the first stage of heat contacts M to H1.

- With the contacts M to H1 closed, the 20 vdc heating fan relay coil (3R) (Line 19), wired from terminal ⟨F2⟩ to ⟨F⟩ on, the sequencer, is energized. Contacts 3R (Line 8) close.
- At the same time, heat relay coil (1HR) (Line 25A), wired from ⟨1⟩ (Line 19) to common on the sequencer, is energized.
- The time delay relay coil (1TR) (Line 18) is also energized.
- After about 10 seconds (minimum) of heating, the contacts in the time-delay relay (1TR) close. This completes the circuit to heat relay coil (2HR) (Line 25), and the time-delay relay coil (2TR)
- Approximately 20 seconds (minimum) after the thermostat closes for heating, the contacts in the time-delay relay (2TR) close, completing a circuit to heat relay coil (3TR) (Line 24).
  This sequencer is staged for two stages of heat with (1HR), (2HR) and (3HR) on the first stage of heat H1. Heaters (4HR), (5HR) and (6HR) are on the second stage of heat H2. The time-delay relay (3TR) is removed for two-stage heating operation. A jumper wire (not shown) is run from ⟨H2⟩ on the sequencer to the socket from which the (3TR) relay was removed.
- If the temperature around the thermostat drops another 1-1/2 to 2 °F, contacts M to H2 close and heat relay coil (4HR) (Line 23) is energized. At the same time, time delay relay coil (4TR) is energized.
- Approximately 10 seconds (minimum) later, the time delay relay contacts (4TR) close, energizing heater relay coil (5HR), and time delay relay coil (5TR).
- Approximately 10 seconds (minimum) later, the time-delay relay contacts (5TR) close, energizing heater relay coil (6HR).
- When the temperature around the thermostat rises 1-1/2 to 2 °F, the second stage heat contacts open, breaking contacts M to H2.
- With the contacts M to H2 open, the heater relay coils de-energize, in 10 second (minimum) intervals, in the same order in which they were energized, (4HR), (5HR), (6HR).
- When the temperature around the thermostat rises to the setting on the thermostat, contacts M to H2 open, de-energizing the heater relay coils, (1HR), (2HR), (3HR), in 10 second (minimum) intervals, in the same order in which they were energized.
- The system has *fan logic*, which means that the fan relay coil (3R) (Line 19), remains energized until all heater relay coils are de-energized.
- Temperature limit switches 2TLS and 1TLS (Lines 17,18), which are automatically reset, are mounted in the heating package, and open one leg of the 24 vac circuit to the sequencer, de-energizing the 20 vdc circuit, cutting all power to the heating relays, time-delay relays and the fan relay.

Recheck the low voltage, 20 vdc circuit, until you understand the sequence. Then proceed to the high voltage circuit.

Set thermostat to HEAT, fan on AUTO, and raise the temperature setting to close contacts M to H1.

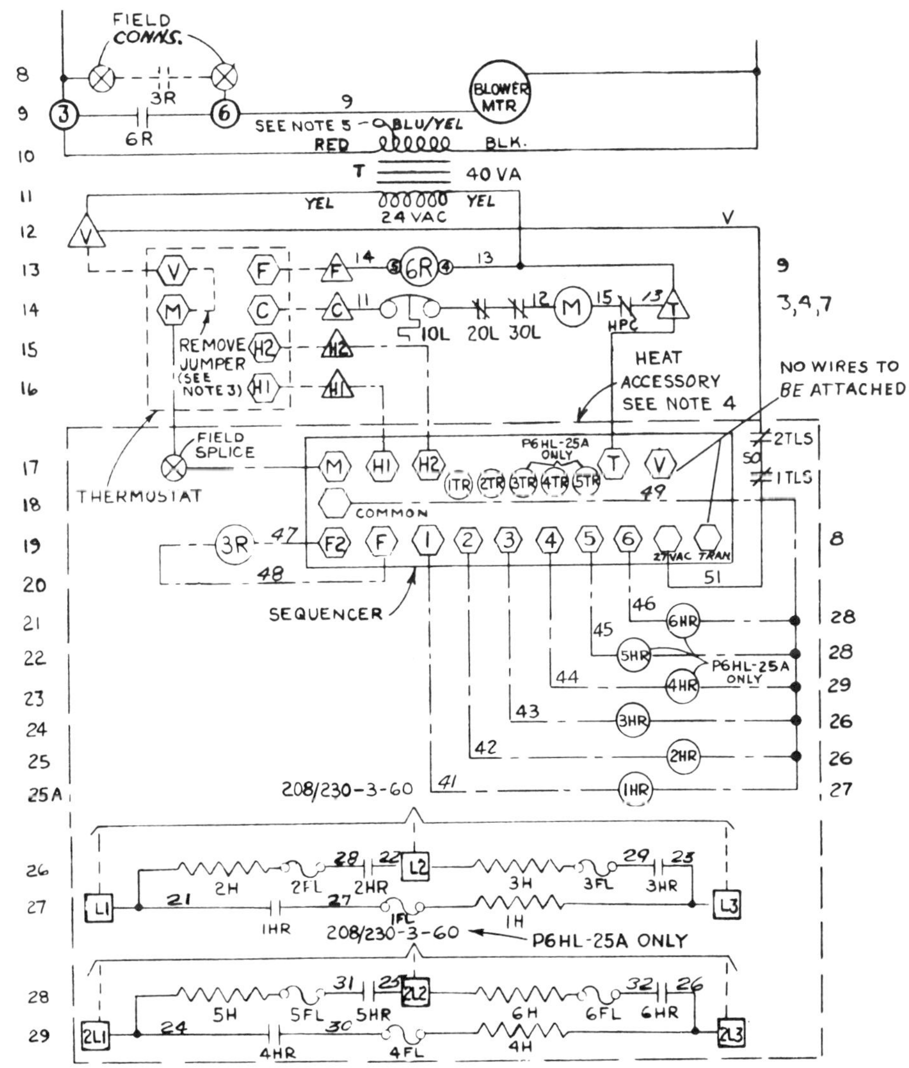

Courtesy, *York*

LINE VOLTAGE

- With contacts M to H1 closed, the heating fan relay coil (3R) (Line 19), is energized, closing contacts 3R (Line 8) and completing a circuit to the blower motor (BLOWER MTR) from L1 to L3. Relay 3R is field installed and bypasses cooling fan contacts 6R.
- At the same time, heater relay coil 1HR is energized, closing contacts 1HR (Line 27), and energizing heater element 1H between L1 and L3.
- Heater relay coil 2HR closes contacts 2HR (Line 26), energizing heater element 2H between L1 and L2.
- Heater relay coil 3HR closes contacts 3HR (Line 26), energizing heater element 3H between L2 and L3.
- If the temperature around the thermostat drops another 1-1/2 to 2 °F, the second stage contacts M to H2 close, energizing heater relay coil 4HR.
- With heater relay coil 4HR energized, contacts 4HR (Line 29) close, completing a circuit to heater element 4H from 2L1 to 2L3.
- Heater relay coil 5HR closes contacts 5HR (Line 28) energizing heater element 5H between 2L1 to 2L2.
- Heater relay coil 6HR closes contacts 6HR (Line 28), energizing heater element 6H between 2L2 to 2L3.

Examining the schematic:

- All heating elements are protected by a fusible link set to open at 333 °F.
- If Line L1 opens, heater 3H will still heat.
- If Line L2 opens, heater 1H will still heat.
- If Line L3 opens, heater 2H will still heat.
- If Line 2L1 opens, heater 6H will still heat.
- If Line 2L2 opens, heater 4H will still heat.
- If Line 2L3 opens, heater 5H will still heat.

Remember, when troubleshooting 3-phase equipment, check the main voltage to terminals L1, L2, L3, 2L1, 2L2, and 2L3 first.

Set the thermostat on HEAT, fan on AUTO, and raise the temperature setting to close contacts M to H1 and M to H2. Wait approximately 60 seconds for all heater elements to energize. Now, what voltages would you read across the following:

1. Coils 1HR, 2HR, 3HR, 4HR, 5HR, 6HR (Lines 25A, 25, 24, 23, 22, 21)
2. Coil (3R) (Line 19)
3. Heater element 3H (Line 26)
4. Heater element 5H (Line 28)
5. Contacts 2TLS (Line 17)
6. Contacts 1TLS (Line 18)
7. Contacts 2HR (Line 26)
8. Contacts 6HR (Line 28)
9. Contacts 3R (Line 8)
10. Blower Motor Terminals (Lines 8, 9)

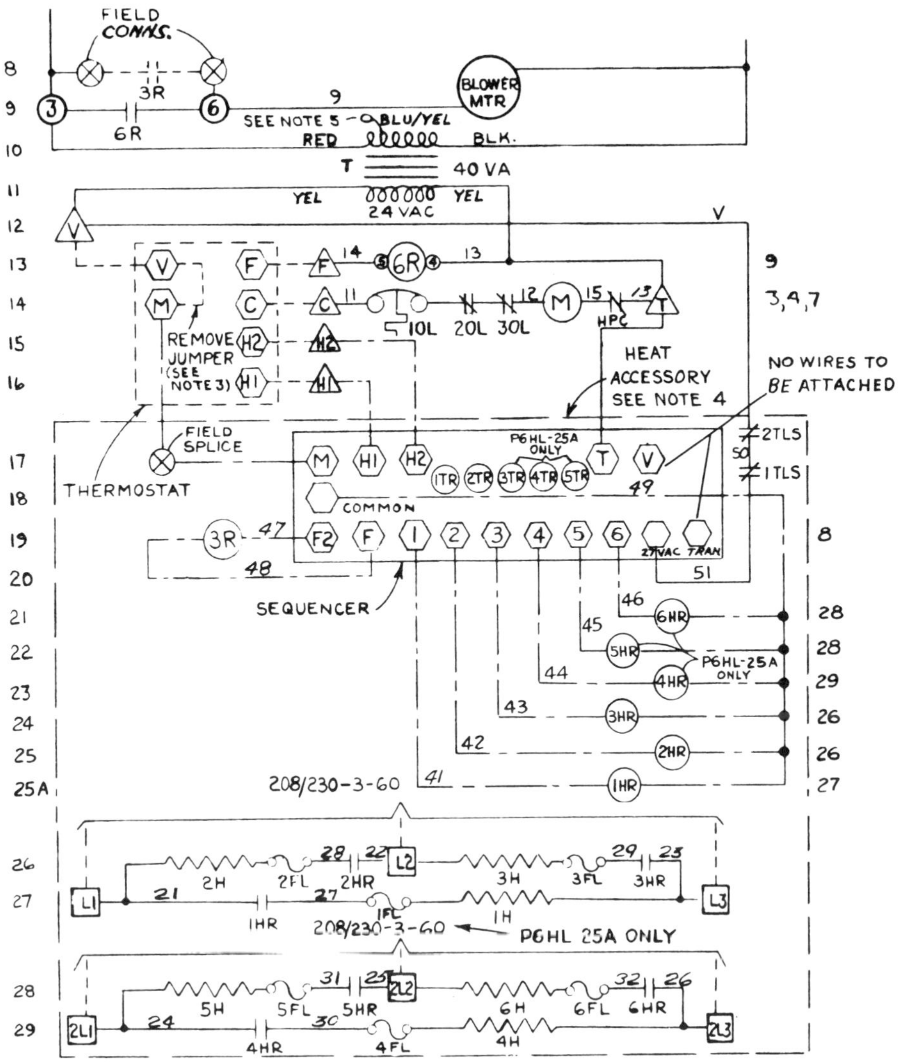

Courtesy, *York*

Using the five-step procedure, solve the following problems:

Problem 1. Heater elements 4H, 5H, 6H do not heat. Heater elements 1H, 2H, 3H and the blower motor are working.

Problem 2. Heater elements 2H, 3H do not heat. Heater elements 1H, 4H, 5H, 6H and the blower motor are working.

Problem 3. Blower motor does not come on. All heaters are energized.

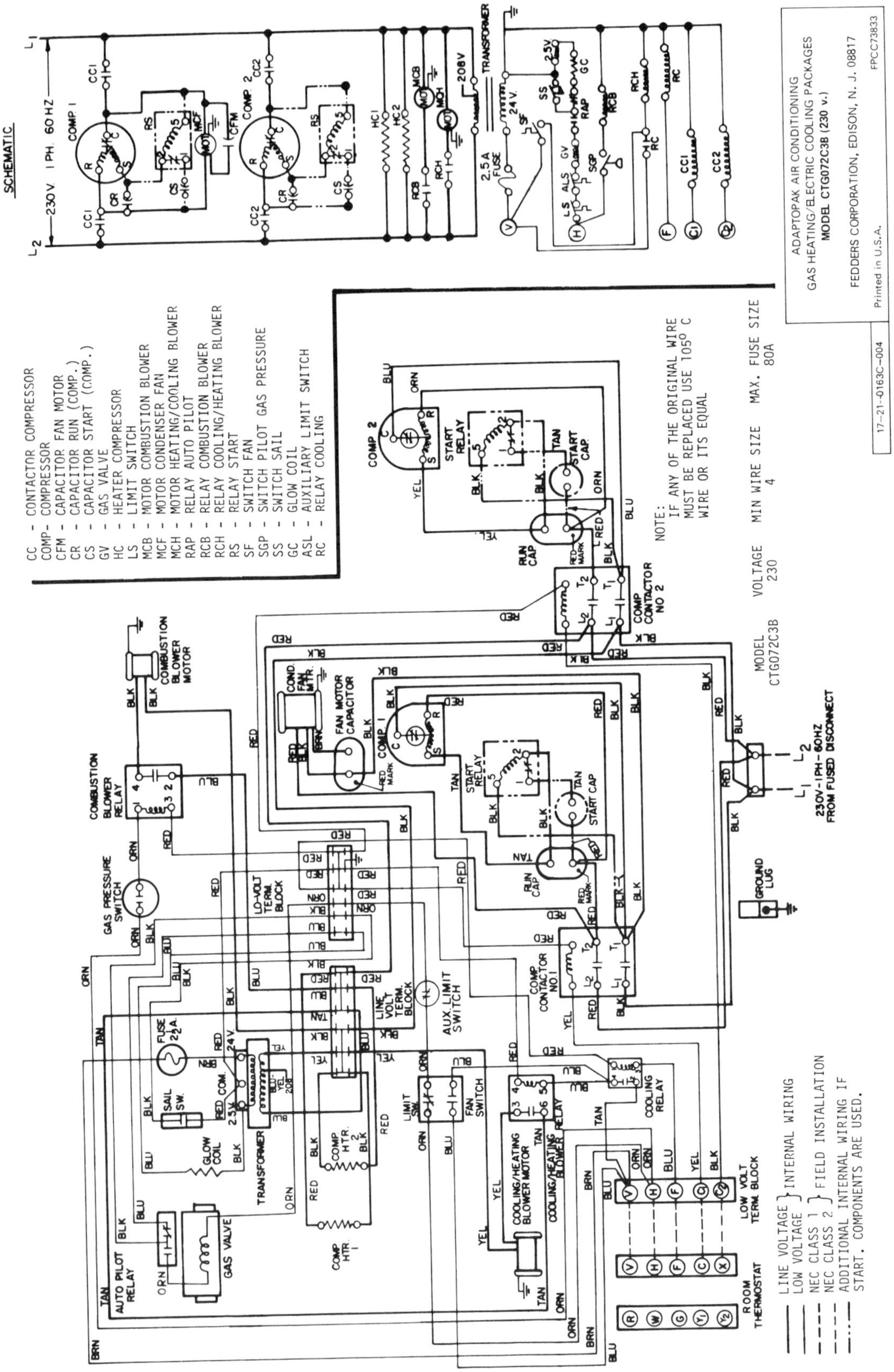

Figure 6-13. Courtesy, *Fedders*

FEDDERS ROOFTOP GAS HEATING AND ELECTRIC COOLING PACKAGE

This commercial-size, 72,000 Btu rooftop package, Figure 6-13, will automatically heat or cool from a field-wired, low voltage, two-stage cooling, one-stage heating thermostat. The schematic does not show the low-voltage thermostat, but it does indicate the terminals where the low-voltage wires are field connected. Terminal (V) is to the fused side of the transformer secondary, terminal (H) is to the heating control circuit, terminal (F) is to the evaporator blower, terminals (C1) and (C2) are to the control circuits for compressor 1 and compressor 2.

1. GO STEP-BY-STEP THROUGH THE ELECTRICAL SEQUENCE

COOLING CYCLE

Set the thermostat on COOL, fan on AUTO. Lower the thermostat setting to energize the first stage, but not the second stage, of cooling. 230 V will be present at L_1 and L_2.

- With the thermostat closed, first stage only, a circuit is completed from the 24-V secondary of the transformer 24V through the fuse to (V) and to (C1). 24 V then energizes the compressor contactor coil CC1 and completes the circuit back to the other side of the transformer secondary.
- If the temperature at the thermostat rises another 1-1/2 to 2 °F, the second stage of the thermostat will close, completing a circuit from (V) to (C2). 24 V will energize the compressor contactor coil CC2 and complete the circuit to the other side of the transformer secondary.
- When the thermostat calls for cooling, a circuit from (V) is automatically completed through the thermostat fan selector to (F). 24 V will energize the cooling relay coil RC and complete the circuit to the transformer secondary.
- With coil RC energized, the normally-open contacts RC in the 24-V circuit will close, energizing the cooling-heating blower relay coil RCH from (V), and completing this circuit to the transformer secondary.

From this information, we can see that if coil CC1 is defective, coil CC2 and coil RC can still function. With parallel circuits, each circuit operates independently and is not affected by the defects in the others. In the schematic, it can be seen that the coils CC1, CC2 and RC are controlled only by the thermostat.

Further observation will show that the blower relay coil RCH is controlled by the closing of contacts RC. However, coil RCH can also be energized by the closing of the temperature-activated fan switch SF, on a rise in temperature, thereby completing a circuit from (V) that bypasses the fan switch. The fan switch is closed in the heating cycle to start the blower motor. It can be seen that a defective cooling relay coil RC will not start the blower on the cooling cycle, but the blower could operate during the heating cycle.

Recheck the low-voltage circuit and proceed to the high-voltage section of the schematic.

- From L_1, a wire is connected to the line side of one set of normally-open contacts CC1 on the compressor contactor. L_2 is connected to the other set of normally-open contacts CC1 on the compressor contactor.
- From the schematic, it can be seen that compressors 1 and 2 have internally-protected C, split-phase, capacitor-start CS capacitor-run CR motors, using a potential start relay covered in Section II.
- With coil CC1 energized, contacts CC1 and CC1 close, starting compressor 1, through the start relay RS.
- When contacts CC1 and CC1 close, they also energize the condenser fan motor MCF The schematic shows the condenser fan motor as a permanent split capacitor motor, showing only a run capacitor CFM and no start relay.
- On a call for the second stage of cooling, C2 is also closed, energizing coil CC2, closing the normally-open contacts CC2 and CC2, starting compressor 2.
- Compressor heaters HC1 and HC2, also called crankcase heaters, do not have switches for operation. They are connected hot from L_1 to L_2.
- With contacts RC closed, the blower relay coil RCH in the 24-V circuit is energized, closing the normally-open contacts RCH in the high voltage circuit, completing a circuit from L_1 through the blower motor MCH to L_2, starting the evaporator blower.

As an electrical troubleshooter, it is important to take notice of the condenser fan motor MCF circuit. The schematic shows that the condenser fan motor starts with the first stage of cooling. The MCF motor is parallel to compressor 1 and in series with the compressor contactor contacts CC1 and CC1. If the CC1 coil in the low-voltage circuit fails and the CC2 coil is good, compressor 2 will operate, but compressor 1 and the condenser fan motor will not operate, causing excessive compressor pressure. The schematic does not show a high-pressure safety control that would shut down the compressor when the compressor pressure goes up. In this system, compressor 2 will cycle on and off, as the internal protector C in the compressor opens and closes.

- The step-down transformer used has either a 230 or 208-V primary connected hot from L_1 to L_2. The secondary has both a 24-V lead for operation of all low-voltage controls and a 2.5-V lead to operate the glow coil GC in the heating circuit.

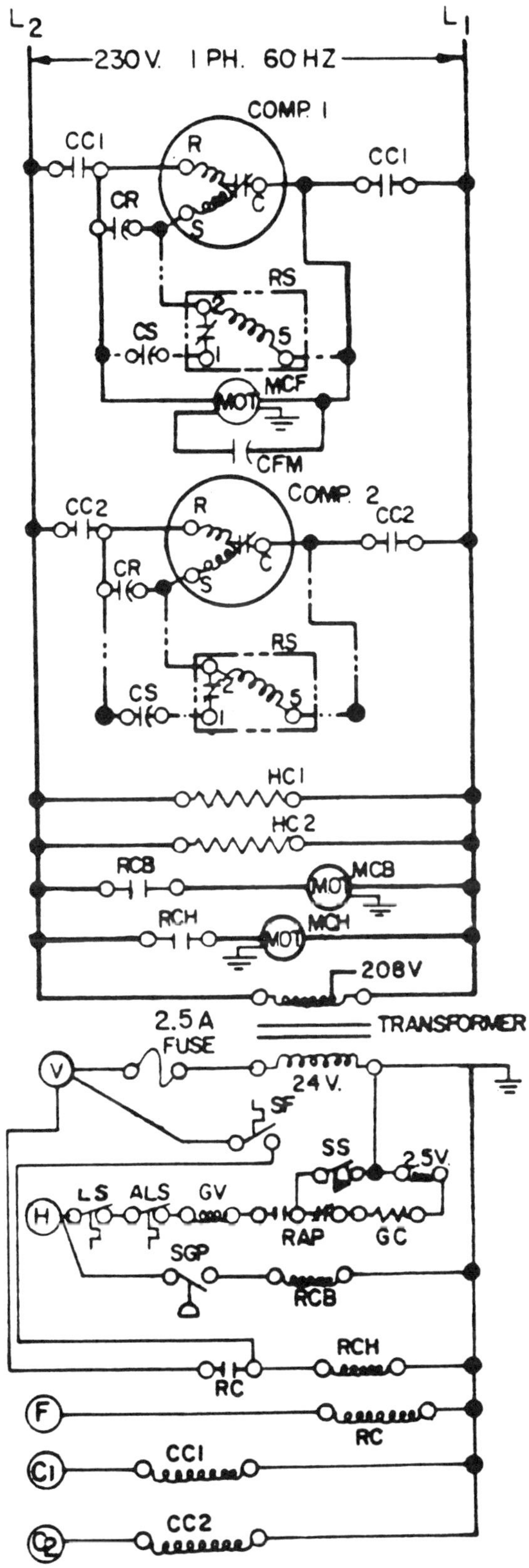

Courtesy, *Fedders*

This completes the cooling side of the schematic. With the thermostat in the COOL position, fan on AUTO and both stages of cooling closed, what voltages would be read, on a voltmeter, across the following switches, coils and motors:

1. Coil CC1
2. Coil CC2
3. Coil RC
4. Coil RCH
5. Coil HC1
6. Contacts RC
7. Contacts RCH
8. Contacts CC1
9. Contacts CC2
10. Motor MCF
11. Motor MCH

Using the 5-step procedure, solve these three problems:

Problem 1. Second stage compressor 2 does not come on. Compressor 1 and the condenser fan are operating. The evaporator blower is working.

Problem 2. Compressor 1 does not run. Compressor 2 does operate. Condenser fan operates. Evaporator blower is working.

Problem 3. The cooling blower is not operating. Compressors 1 and 2 and the condenser fan are operating.

Having covered the cooling side of the schematic, now check the heating circuit. This heater is fired by natural gas. Due to high winds found on roof tops, a pilot flame could be extinguished easily. To solve this problem, a glow coil is used to automatically ignite the pilot whenever it is extinguished. Combustion air for the burners is supplied by the combustion blower, which must operate before the gas valve will open.

A closer look at the wiring diagram reveals that the transformer has a tapped secondary. This is a dual-voltage secondary with a common center tap. From the red common to the brown lead is 24 V, and from the red common to the black lead is 2.5 V. The 2.5 V energizes the glow coil. A mistake in wiring at this point will cause the glow coil to burn out.

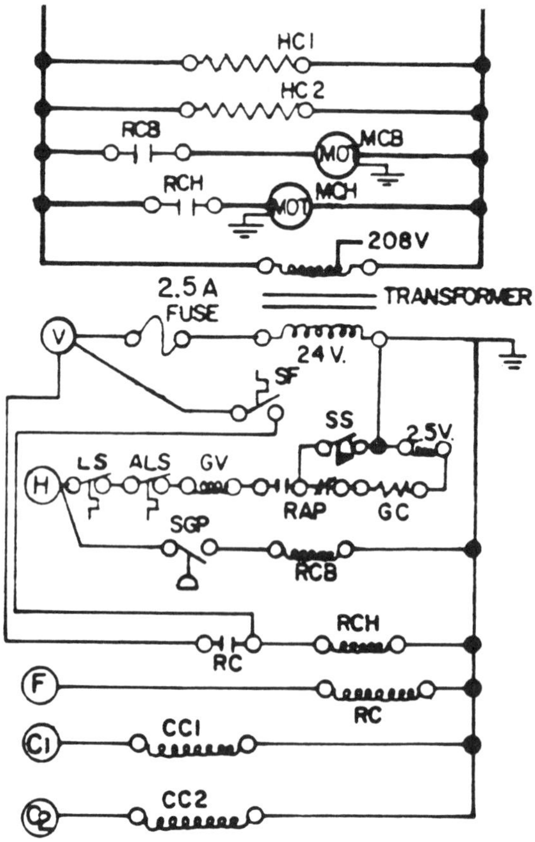

HEATING CYCLE

Place the thermostat on HEAT, fan on AUTO, and raise the thermostat setting to close the heating contacts. 230 V is present at L_1 and L_2.

- With the thermostat closed and calling for heat, a circuit is completed from (V) to (H), energizing the combustion blower relay coil, if the gas pressure switch is closed. The gas pressure switch is connected to the main gas line and is closed if there is gas in the main line.
- Air velocity from the combustion blower will close the sail switch, completing a circuit from the 2.5-V transformer tap to the glow coil through the normally-closed half of the auto pilot relay.
- The auto pilot relay has one set each of normally-open and normally-closed contacts. The pilot flame heats a thermocouple, which opens the normally-closed contacts, breaking the circuit to the glow coil. Simultaneously, the normally-open contacts close, completing the 24-V circuit to the gas valve coil, through the normally-closed limit switch and the normally-closed auxiliary limit switch, when the sail switch is closed.
- Whenever the pilot flame goes out, the thermocouple cools, returning the auto pilot relay contacts to their normal positions. The glow coil will not energize, again, unless the combustion blower motor is running, closing the sail switch. When the auto pilot relay is in the normal position, the gas valve is de-energized, shutting off the flow of gas to the burners.

- As the temperature in the heat exchanger rises to approximately 140 °F, the fan switch SF closes and, by bypassing contacts RC, completes a circuit from (V) to the cooling-heating blower motor relay coil RCH

Recheck the low voltage circuit before tracing the high-voltage circuit.

- The energized coil RCB in the 24-V circuit closes contacts RCB in the 230-V circuit, starting the combustion blower motor MCB.
- With the blower motor relay coil RCH energized, contacts RCH in the 230-V circuit close, starting the cooling-heating blower motor MCH.

Observe that if the heating contacts (V) to (H) in the thermostat open, the gas valve coil GV is de-energized, closing the flow of gas to the main burners.

Opening contacts (V) to (H) also breaks the circuit to the combustion blower relay coil RCB, opening contacts RCB, stopping the combustion motor MCB. As the combustion motor slows down, the sail switch SS opens, preventing the gas valve GV from opening and the glow coil GC from operating until the combustion motor is running.

Until the temperature in the heat exchanger drops to approximately 120 °F, the fan switch SF will not open, shutting off the blower motor MCH,

Further observation will show that the combustion blower motor will continue to operate when either the limit switch LS or the auxiliary limit switch ALS is open, breaking the circuit to the gas valve coil.

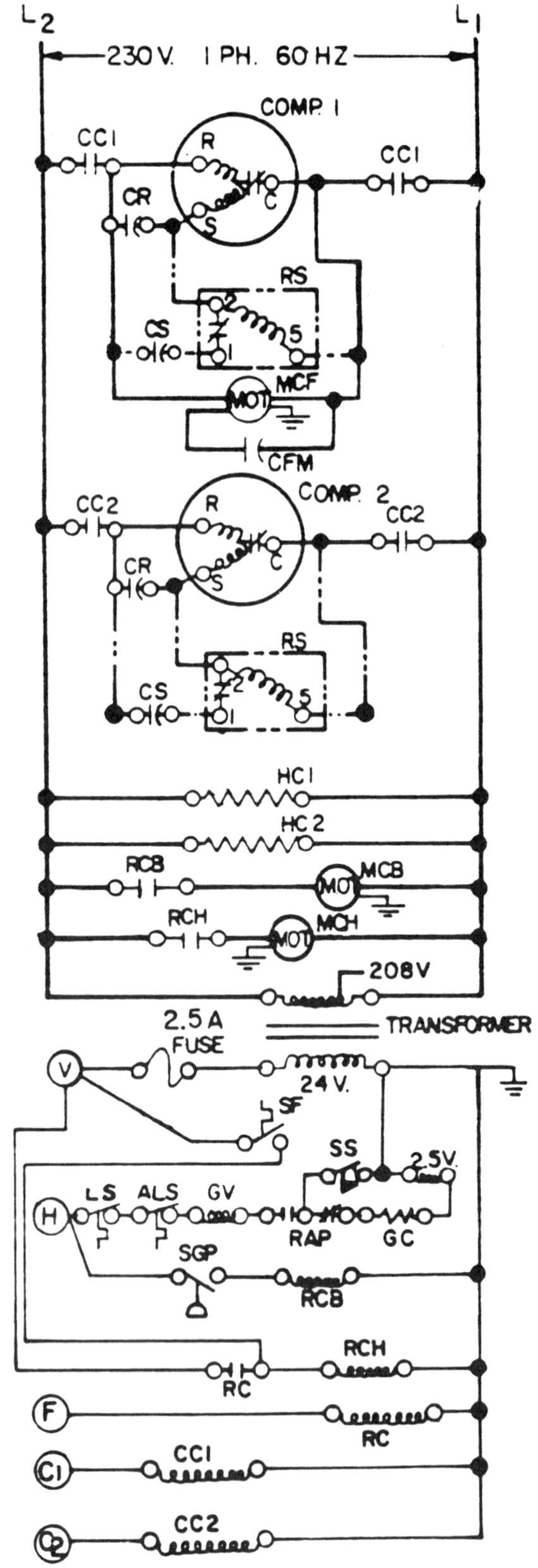

Courtesy, *Fedders*

With the thermostat in the HEAT position, fan on AUTO, the thermostat contacts closed for heat, and the burners firing, what voltages would be read on a voltmeter across the following coils, switches and motors:

1. Coil RCB
2. Coil RCH
3. Coil GV
4. Coil GC
5. Switch SGP
6. Switch SS
7. Switch SF
8. Contacts RCB
9. Contacts RCH
10. Motor MCB
11. Motor MCH

Problem 1. Burner will not ignite. Pilot is on.

Problem 2. Heating-cooling blower motor will not run on heating, but will operate on cooling and constant fan position on thermostat.

Problem 3. The pilot is out and will not ignite automatically from glow coil.

WILLIAMSON FIVE-IN-ONE OIL UNIT

The Williamson Five-in-One, Figure 6-14, is a complete package consisting of an oil-fired, forced-air furnace, an air conditioning cooling coil for cooling and dehumidification, an electronic air cleaner for removing dust and other airborne particles, and an automatic humidifier to add moisture to the air during the heating season.

The Five-in-One comes factory wired, except for the low-voltage control wire to the outside air conditioner condensing unit, low-voltage humidistat, and low-voltage thermostat. These parts are connected to the Five-in-One by field wiring.

The legend accompanying the wiring diagram is different from that used by other manufacturers. Williamson uses a solid line ____ for 115 V, a dashed line with a dot -.-.- for 24 V and a dashed line - - - for dc voltage. The dc voltage is developed in the electronic air cleaner power pack to supply the two cells with 4,000 vdc and 8,000 vdc.

The schematic shows all wires to be of equal weight and solid for 115 V, 24 V and field wiring.

The R856 or R8239 fan center is a Honeywell control, incorporating a 24-V transformer (R856 XFMR) and a fan relay coil (FAN RELAY) with a set of normally-open contacts (R856) to start the indoor fan motor (FAN MOTOR).

The legend on the schematic indicates ○ for terminals on the fan center, □ for thermostat subbase connections, and ◇ for terminals on the primary safety control.

1. GO STEP-BY-STEP THROUGH THE ELECTRICAL SEQUENCE

To understand this schematic, we will cover each operation separately.

COOLING

Place the thermostat selector on COOL (C O H), fan on AUTO (ON AUTO) and lower the temperature setting below room temperature to close the cooling contacts (COOL).

- With the cooling contacts closed, a circuit is completed from (R), one side of the 24-V transformer, through [R] and the thermostat selector, to terminal (Y), then to the compressor contactor coil (COOL CONT.), in the outside condensing unit. The circuit returns to terminal (C), the other side of the transformer secondary.
- When the cooling contacts close, another circuit is completed, through the fan switch (ON AUTO), from [R] to [G]. This energizes the fan relay (FAN RELAY) which is connected between (G) and the return side of the transformer secondary (C).
- With the fan relay coil energized, the normally-open contacts in the fan center close (R856), completing a circuit from the 115-V hot line, through terminal (3) to the fan motor (FAN MOTOR) and, from there, to the GROUND wire.

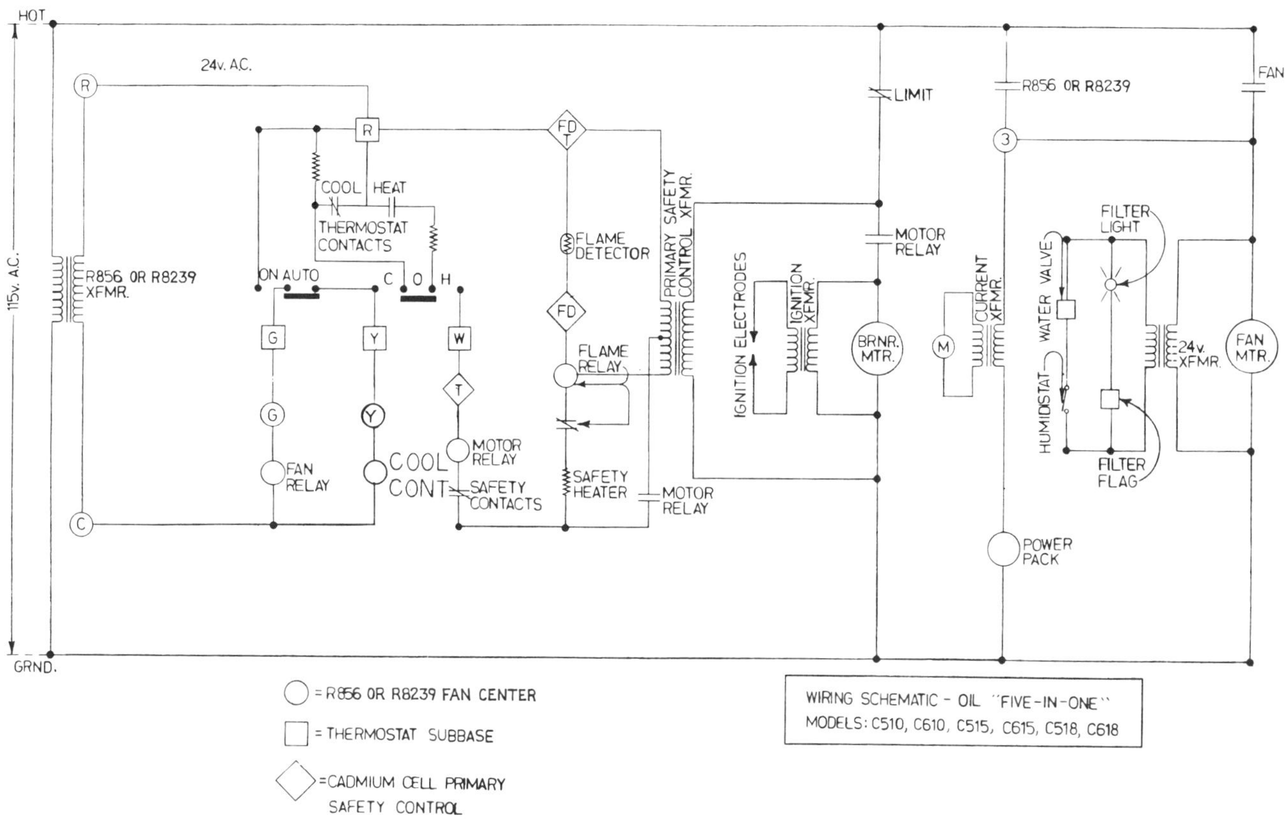

Figure 6-14. Courtesy, *Williamson*

HEATING CYCLE

Place the thermostat selector switch on HEAT (C O H), leave the fan switch on AUTO, and raise the temperature setting above room temperature to close the heating contacts (HEAT).

The low voltage for the heating cycle is supplied by the primary safety control transformer (PRIMARY CONTROL), located in the oil primary control. The primary of the step-down transformer is wired between the HOT and GROUND wires, through the normally-closed contacts of the limit switch (LIMIT). The limit switch is set to open if the discharge air temperature exceeds 200 °F. The opening of the limit switch will break the circuit to the transformer primary, shutting down the oil primary control.

- With the heating contacts closed (HEAT), a circuit is completed from one side of the safety control transformer secondary, through (FD T) and (R), the thermostat (W) and (T), to terminal on the primary control.
- From (T), the circuit continues through the burner motor relay coil (MOTOR RELAY) and the normally-closed safety contacts (SAFETY CONTACTS), to the safety heater (SAFTY HEATER). From the safety heater, the circuit proceeds through the normally-closed flame relay contacts and the flame relay (FLAME RELAY) to the transformer secondary.

- With the burner motor relay coil energized, both sets of normally open motor relay contacts MOTOR RELAY close and the closed motor relay contacts in the 24-V primary control complete a second circuit from the motor relay coil to the center tap PRIMARY CONTROL of the transformer secondary. The center tap supplies only a minimal voltage to keep the relay coil energized. The end tap of the transformer supplies a maximum voltage to energize the coil and pull in the relay.
- The second set of normally-open motor relay contacts in the 115-V circuit also closed, completing a circuit through the limit switch LIMIT to both the burner motor BRNR MTR and the primary of the 10,000-V ignition transformer IGNITION XFMR The motor starts, resulting in a spark at the ignition electrodes If the burner motor is pumping oil to the burner nozzle, there is ignition.
- A flame detector FLAME DETECTOR is wired between terminals FD and FD and connected to the end tap of the primary transformer through the flame relay FLAME RELAY. In the absence of light, the cad cell of the flame detector has a very high resistance and blocks the flow of current to the flame relay.

  However, when light from the oil flame strikes the flame detector, the resistance of the cad cell falls to a very low value and a circuit is completed, through the flame detector, to the flame relay.
- Energizing the flame relay opens the normally-closed flame relay contacts and takes the safety heater SAFETY HEATER, out of the circuit, thereby assuring that the safety contacts SAFETY CONTACTS will not open.

Had the oil failed to ignite, the flame detector would have blocked the current flow to the flame relay and, after 45 seconds, the safety heater would have opened the safety contacts and cut off the entire oil delivery and ignition systems by deactivating the motor relay. The oil burner cannot be restarted, even with the thermostat calling for heat, until the safety heater has cooled and the safety contacts closed.

On restart, only 45 seconds are allowed for ignition. If ignition is not achieved in these few seconds, the safety heater will once again open the safety contacts.

- The normally-open fan contacts FAN are set to close when the temperature in the heat exchanger reaches 110 °F. They have a cutout temperature differential of approximately 25 °F. The closed fan contacts complete the circuit to the fan motor from the HOT to GROUND wires.

  The fan motor is not controlled by the low-voltage control circuit and will continue to run until the temperature in the heat exchanger drops to the cutout temperature, even though the primary control or limit switch has shut down the system on safety. Thus, the fan continues to operate as long as the air it delivers is at a comfortable temperature.
- A 115 to 24-V step-down transformer 24V XFMR is wired parallel to the fan motor FAN MOTOR and is hot whenever the fan motor is energized. Wired across the 24-V secondary is a filter flag switch . The filter flag switch is a normally-open switch that is kept open by the flow of air through the filter. When the filter becomes blocked, restricting the air flow, the flag switch will close and energize the filter light which warns that the filter needs to be replaced.

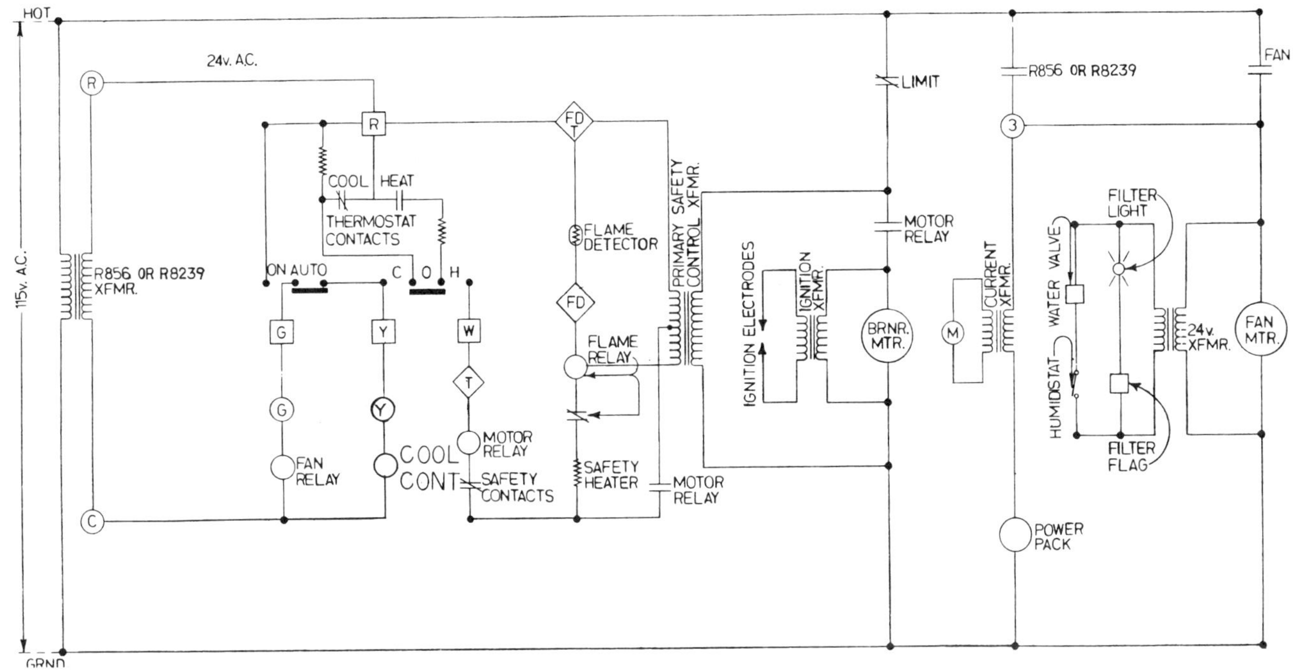

Courtesy, *Williamson*

HUMIDIFIER

The humidifier adds moisture to the return air when the humidistat closes. To prevent the humidifier from operating when the fan is off, the 24-V control transformer is wired parallel to fan motor. The water valve in the humidifier is opened by a 24-V solenoid coil wired in series with the humidistat.

ELECTRONIC AIR CLEANER

The electronic air cleaner will remove and hold airborne particles and should only operate when the fan motor is running.

To make sure that air cleaner operates only when the motor is running, the power pack (which is the complete control circuit for the air cleaner) and the current transformer (that controls the operation of the performance meter (M) mounted on the air cleaner door) are wired in series with the fan center contacts R856 (cooling) and the temperature operated contacts of the fan relay FAN (heating). The closing of either set of contacts will energize the electronic air cleaner through terminal (3).

Set the thermostat to HEAT, fan to AUTO and raise the temperature setting of the thermostat to close R to W. Let the fan start.

Now, what voltages would you read across the following coils, switches and motors:

1. Motor relay coil
2. Flame relay coil
3. Fan motor
4. Burner motor
5. Fan switch
6. Motor relay contacts
7. Flame relay contacts
8. Filter flag
9. Humidistat (open)

Solve these problems using the five-step procedure:

Problem 1. Fan motor will not run in the cooling cycle.

Problem 2. Burner motor and ignition cut out after 45 seconds.

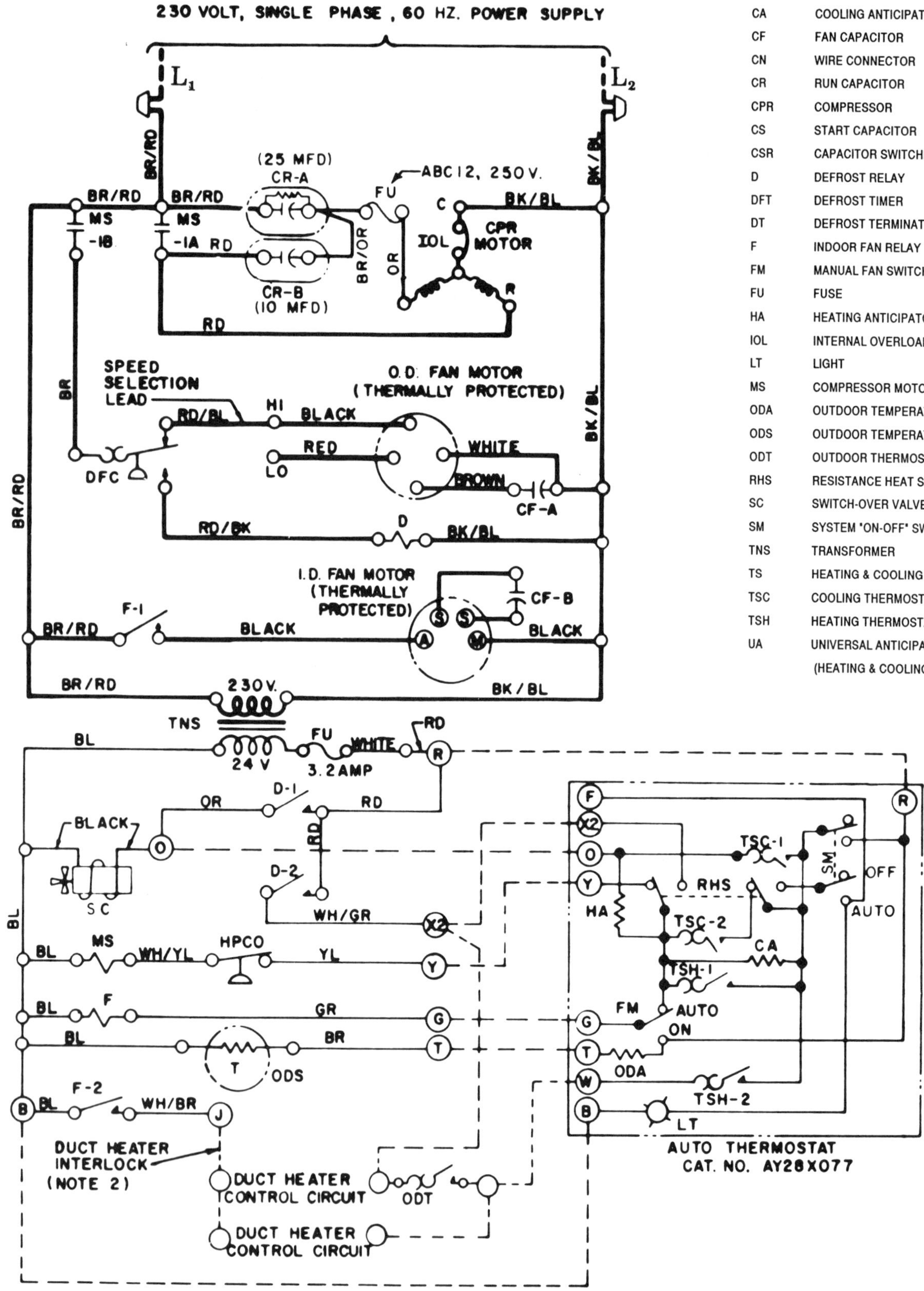

CA	COOLING ANTICIPATOR
CF	FAN CAPACITOR
CN	WIRE CONNECTOR
CR	RUN CAPACITOR
CPR	COMPRESSOR
CS	START CAPACITOR
CSR	CAPACITOR SWITCHING RELAY
D	DEFROST RELAY
DFT	DEFROST TIMER
DT	DEFROST TERMINATION SWITCH
F	INDOOR FAN RELAY
FM	MANUAL FAN SWITCH
FU	FUSE
HA	HEATING ANTICIPATOR
IOL	INTERNAL OVERLOAD PROTECTOR
LT	LIGHT
MS	COMPRESSOR MOTOR CONTACTOR
ODA	OUTDOOR TEMPERATURE ANTICIPATOR
ODS	OUTDOOR TEMPERATURE SENSOR
ODT	OUTDOOR THERMOSTAT
RHS	RESISTANCE HEAT SWITCH
SC	SWITCH-OVER VALVE SOLENOID
SM	SYSTEM "ON-OFF" SWITCH
TNS	TRANSFORMER
TS	HEATING & COOLING THERMOSTAT
TSC	COOLING THERMOSTAT
TSH	HEATING THERMOSTAT
UA	UNIVERSAL ANTICIPATOR (HEATING & COOLING)

Figure 6-15. Courtesy, *General Electric*

GENERAL ELECTRIC SINGLE PACKAGE HEAT PUMP

The heat pump, Figure 6-15, is basically an air conditioner that will cool in summer and heat in winter.

To better understand the operation of the heat pump, imagine it first as an ordinary window air conditioner. The inside coil, or evaporator, extracts heat from the room and expels it, through the outside coil, or condenser, to the outside air.

Now imagine that, in the fall, the window air conditioner is reversed, with the condenser being the inside coil and the evaporator being the outside coil. Your air conditioner will now extract heat from the outside air and expel it into the room.

The heat pump reverses the operation of the air conditioner very simply, by means of a four-way reversing valve. When the valve is shifted, by a change in temperature, the flow of refrigerant is reversed. This, in effect, converts the evaporator into a condenser and the condenser into the evaporator. The automatic operation of this reversing valve is controlled by a wall-mounted thermostat that signals whether heating or cooling is required in the conditioned space.

1. GO STEP-BY-STEP THROUGH THE ELECTRICAL SEQUENCE

COOLING CYCLE

A 24-V thermostat is used for heat pump operation. There are two stages of heat, TSH-1 and TSH-2, and two stages of cooling TSC-1 and TSC-2.

This model thermostat has an AUTO position on the selector switch. The system will automatically switch from cooling to heating or heating to cooling as the temperature at the thermostat changes.

Set the thermostat switch on AUTO, fan on AUTO, and lower the thermostat setting to close the first stage cooling contacts only. 230 V is present at L_1 and L_2.

- With the thermostat contacts closed and the system switch on AUTO, a circuit is completed from the fused side of the 24-V transformer, through (R) (R), to the (O) terminal on the thermostat subbase.
- Field wiring — — — connects (O), to the (O) terminal on the outside unit. The circuit from (O) energizes the solenoid coil on the reversing or switch-over valve, changing the valve's position to cooling. The circuit is completed to the other side of the transformer secondary.
- Should the temperature around the thermostat rise another 1-1/2 °F, the second stage cooling contacts close, making a circuit to (Y) on the subbase.
- Field wiring — — — connects (Y) to the (Y) terminal in the outside unit. The circuit from (Y) energizes the compressor motor contactor coil through the high-pressure cutout, completing the circuit back to the other side of the transformer.

- When the second stage contact in the thermostat closes, a circuit is also initiated to terminal (G) on the subbase, through the fan switch, which is set in the AUTO position.
- Field wiring connects (G) on the subbase to terminal (G) in the outside unit, energizing the indoor fan relay coil and completing the circuit back to the transformer.

At this point, note that the first stage of cooling only energizes the switch-over valve solenoid. This indicates that the switch-over valve is normally in the heating position. If the switch-over valve solenoid failed while in the cooling cycle, the switch-over valve would change back to the heating position. Note, too, that the second stage in the thermostat is the only control for the compressor motor contactor coil and the indoor fan relay coil.

Now, trace the high voltage cooling circuit.

- With the main disconnect in the ON position, 230 V is supplied to the high side of the step-down control transformer.
- A trickle circuit is hot from the line side of the compressor motor contactor, through the resistor in the run capacitor and the fuse to the start terminal on the compressor. This trickle circuit provides approximately 40 watts of continuous heat. The circuit continues through the internal overload protector to the common terminal C on the compressor. The circuit is completed from C to Line L_2.
- The normally-open contacts and close when the compressor motor contactor coil in the 24-V circuit is energized, .
- With contacts closed, a circuit is completed through the run terminal of the compressor to the common C, terminating at L_2. Another circuit is completed, through the run capacitor and the fuse to the start winding of this PSC motor. The circuit continues through the internal protector to the common terminal C and L_2.
- The other set of contacts in the compressor motor contactor are also closed. This completes a circuit through the defrost control, which is used to defrost the outside coil during the heating cycle. From the defrost control, the circuit continues to the high speed terminal of the outdoor fan motor. The outdoor fan motor is a two-speed permanent split capacitor (PSC) motor. The circuit is completed from the motor to L_2.
- When in the thermostat closes, the indoor fan relay coil is energized, closing the normally-open contacts in the 230-V circuit.
- With the contacts of the indoor fan motor relay closed, a circuit is completed to the indoor fan motor and L_2. The schematic shows the indoor fan motor to be a permanent split capacitor type.

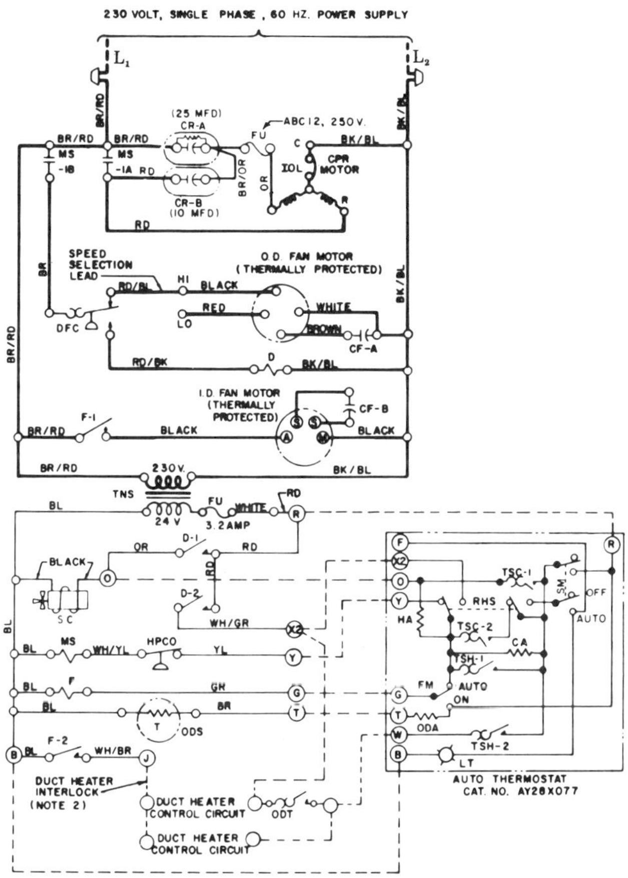

Courtesy, *General Electric*

Note that the compressor motor contactor controls the compressor with one set of contacts MS -1A and the outdoor fan motor (O.D.) with another set of contacts MS -1B. When the compressor motor contactor contacts are open, trickle voltage is still present at the start and common terminals of the compressor. Unlike previous examples, these contacts only break the L_1 line to the compressor and the outdoor fan motor, rather than both L_1 and L_2.

With the cooling side of the schematic completed, set the thermostat selector switch to AUTO and the fan switch on AUTO. Lower the thermostat setting to close both stages of cooling.

What voltages would be read, on a voltmeter, across the following:

1. Coil MS
2. Coil F
3. Coil
4. Switch HPCO
5. Switch F-1
6. Switch DFC
7. I.D. Fan Motor A to M
8. CPR Motor C to R
9. O.D. Fan Motor HI to WHITE

Using the five-step procedure, solve these three problems:

Problem 1. Hot air coming out of supply ducts in the summer. The compressor and outdoor and indoor fans are operating.

Problem 2. The compressor and outdoor fan are not operating. The indoor fan is operating.

Problem 3. The outdoor fan is not running. The compressor and indoor fan are both operating with the compressor cycling on the high pressure cutout.

HEATING CYCLE

To check the heating cycle, place the system selector switch on AUTO, the fan switch on AUTO and move the control arm to close the contacts to the first stage of heat.

- With the thermostat contacts TSH-1 closed, a circuit is run from (R) to (R) through the selector switch and the contacts to (Y) on the subbase.
- (Y), on the subbase is field wired — — — to the (Y) terminal on the heat pump. From (Y), the circuit continues through the high-pressure cutout HPCO to the compressor motor contactor coil MS and terminates at the transformer. With the coil energized, contacts MS-1A and MS-1B, in the 230-V circuit close, starting the compressor and the outdoor fan motor.
- When the first stage heating contacts close, another circuit is completed through the heat anticipator HA to (O). Although (O) is field wired — — — to terminal (O) on the heat pump, the switch-over valve coil is not energized, due to the high resistance of the heat anticipator. Voltage drop through the resistor prevents the valve from shifting out of the heating position.
- Closing of the first stage heating contacts TSH-1 energizes a third circuit through the indoor fan relay coil F. The energized indoor fan relay coil closes contacts F-1 starting the indoor fan motor.

At this point, the heating system is operating on the compressor only. However, as outside temperatures drop, the heat pump may not be able to extract enough heat from the outside air. To supplement the heat output of the pump, electric resistance heaters are installed in the supply duct and are controlled by the second stage of heat TSH-2 in the thermostat.

- If the temperature at the thermostat drops 1-1/2 °F below the thermostat setting, contacts TSH-2 close and complete a circuit from (R), through the selector switch, to (W) on the subbase. Field wiring connects (W) to the normally-open outdoor temperature switch ODT and completes a circuit through the first of two duct heater control circuits to (J).
- When the indoor fan relay coil F is energized, it closes contacts F-1, starting the fan motor and contacts F-2, completing the circuit from the first duct heater control circuit to (B) and the return side of the transformer. Having the two switches, F-l and F-2, in tandem ensures that the fan is operating while the resistance heaters are energized. (Because the heaters themselves are accessories, they are not shown in this schematic.)
- The outdoor thermostat is set to close at an outdoor temperature where additional electric resistance heat will be required. This temperature setting is related to the heat loss of the structure and the capacity of the resistance heating units. When the outdoor thermostat closes, the circuit to the second duct heater control circuit is completed, through (J) and F-2 to (B).

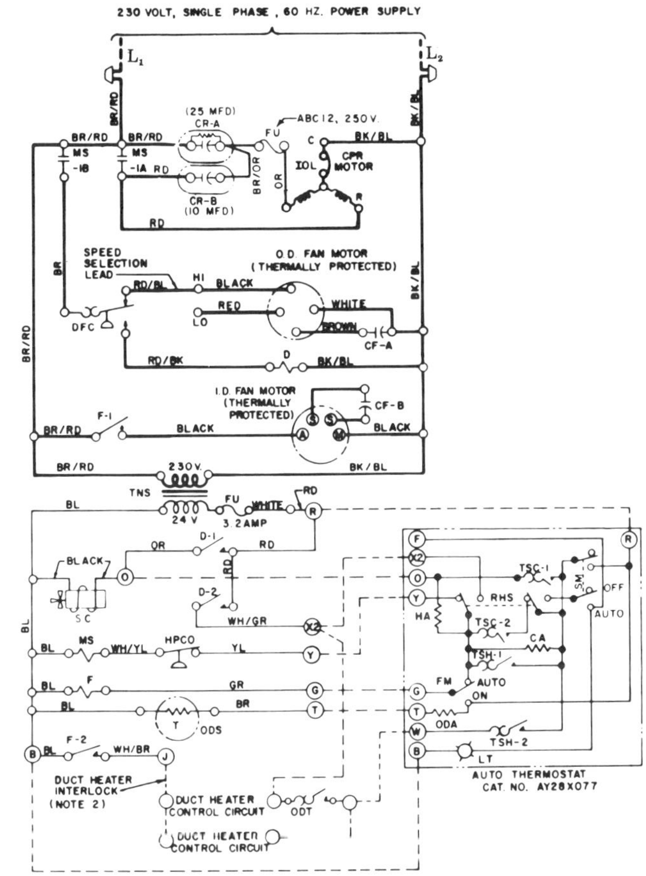

Courtesy, *General Electric*

- In the event of a failure in the heat pump, emergency heat may be required until the heat pump can be repaired. At this time, the emergency heat switch RHS in the thermostat is manually shifted to the ON position.
- With the emergency heat switch in the ON position, the circuits from Ⓡ, through heating contacts TSH-1 to Ⓨ are opened. This de-energizes the compressor motor contactor coil MS, and shuts off the compressor motor and outdoor fan motor by opening contacts MS-1A and MS-1B in the 230-V circuit.
- When the energy heat switch is in the ON position, a circuit is run from Ⓡ through heating contacts TSH-1 to terminal X2 on the subbase.
- Field wiring — — — connects X2 on the subbase to X2 in the outside unit. Another field wire is run to the duct heater control circuit, bypassing the outdoor thermostat ODT, and energizing the second duct heater regardless of the outdoor temperature.
- If the temperature around the thermostat drops 1-1/2 °F below the thermostat setting, heating contacts TSH-2 will close, energizing the first duct heater.
- Heating contacts TSH-1 will already be closed to keep the indoor fan motor running by energizing the coil F. The indoor fan motor will continue to run until the first stage of heat is satisfied.
- When the emergency heat switch is shifted to ON, a light LT on the thermostat comes on through a circuit from Ⓑ on the subbase to Ⓑ in the outside unit. This circuit is completed through the ON contacts of the emergency heat switch and the AUTO contacts of the thermostat switch.

Observing the heating cycle, we can determine:

- The normal position of the switch-over valve is in the heating position.
- A resistance-free circuit must be made from Ⓡ to Ⓞ to reverse the switch-over valve position for cooling.
- The indoor fan relay coil must be energized, closing contact F-2 from Ⓙ to Ⓑ before the duct heaters will energize.
- The emergency resistance heat switch will de-energize the refrigerant system and energize the supplementary resistance heater circuit.
- If the high pressure cutout HPCO opens, both the compressor and the outdoor fan will stop and the other systems will remain operative.

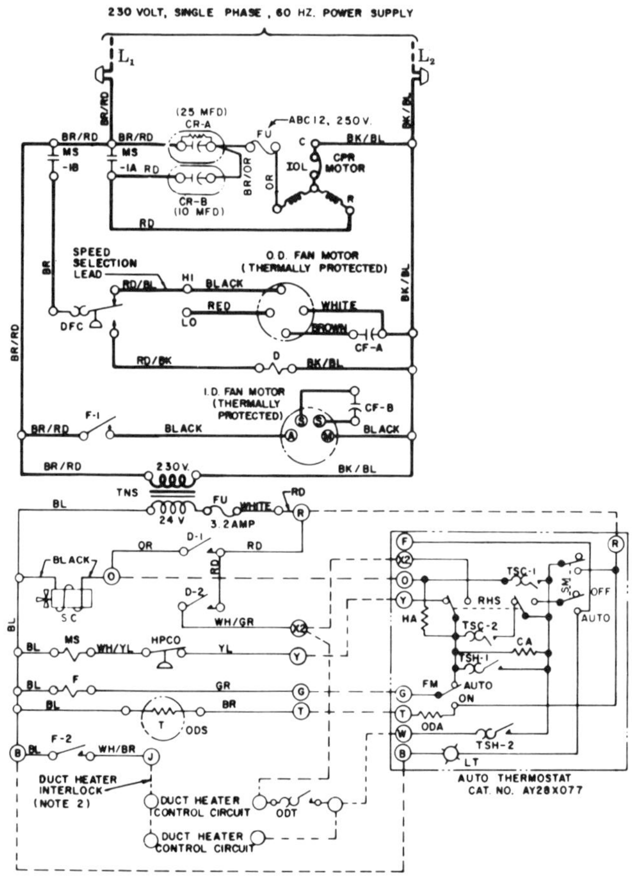

Courtesy, *General Electric*

DEFROST CYCLE

Periodically, during the heating cycle, the outdoor coil will require defrosting. This is accomplished automatically whenever the thermostatic bellows in the defrost control senses below-freezing temperatures and the pressure sensor in the defrost control indicates that air flow through the coil is blocked by ice. The system is in the heating cycle with contacts TSH-1 closed.

- When icing occurs, the defrost control, which is a single-pole, double-throw switch, breaks the circuit to the outdoor fan motor (O.D.), stopping the outdoor fan.
- At the same time, the defrost control completes a circuit to energize the defrost relay coil in the 230-V circuit.
- With the defrost relay coil energized, contacts D-1 and D-2 in the 24-V circuit, close.
- When closed, contact D-1 completes a circuit from (R) to (O), energizing the switch-over valve solenoid and changing the switch-over valve to the cooling position, bypassing the thermostat.
- When closed, contact D-2 completes a circuit from (R) on the transformer to (X2) and then to the duct heater control, bypassing the thermostat.
- The energized electric resistance heater is used to temper or warm the cool air now coming from the indoor coil.
- As the ice on the outdoor coil melts, the static pressure difference is reduced and the outdoor coil temperature reaches between 50 to 60 °F, the defrost control will change position, restarting the outdoor fan, de-energizing the defrost relay coil, opening contacts D-1 and D-2.
- With defrost relay contacts D-1 and D-2 open, the switch-over valve solenoid is de-energized, returning the switch-over valve to the heating position, and opening the circuit to the duct heater.

THE ANTICIPATOR CIRCUITS

The function of the heating anticipator and the cooling anticipator was covered in Section II along with the thermostat. Note that the cooling anticipator is energized whenever the system switch or thermostat is on AUTO. The heating anticipator HA is energized when the thermostat is on AUTO and the first stage heating TSH-1 switch is closed.

A third anticipator circuit ODA is energized whenever line voltage is available, even with the thermostat off. Its purpose is to furnish a variable amount of heat, depending on the outdoor temperature.

The outdoor anticipator ODA located in the thermostat, is connected in series with an outdoor sensor ODS, actually a thermistor, that is exposed to the outdoor air. The thermistor has the ability to vary its resistance with a rise or fall in temperature. In this case, the resistance of the thermistor increases with a temperature drop.

With colder outdoor temperatures, more resistance heat is generated at the thermistor than at the thermostat heat anticipator. Thus, the thermostat senses less *false heat* and the heating system runs longer on each cycle.

At warmer outdoor temperatures, the thermistor has less resistance and the thermostat anticipator generates more heat, thereby fooling the thermostat and foreshortening the heating system running time on each cycle.

With the heating and defrosting schematic completed, set the thermostat selector switch to AUTO, the fan to AUTO and raise the thermostat to close the first stage of heat. What voltages would you read across the following:

1. Coil MS
2. Coil F
3. Coil D
4. Contact DFC to HI
5. Contacts F-1
6. Contacts F-2
7. Contacts MS -1A
8. Contacts MS -1B

Using the five-step procedure, solve these three problems:

Problem 1. Compressor is not running. Outdoor and indoor fans are running.

Problem 2. Indoor fan, compressor and outdoor fan are not running.

Problem 3. One duct heater will not energize. Second stage contacts TSH-2 are closed. Compressor, outdoor fan and indoor fan are all operating.

Section VII

SOLUTIONS

Section 5 Solutions

EXAMPLE 1

1 - 24 V
2 - 0 V
3 - 0 V
4 - 24 V
5 - 24 V
6 - 240 V
7 - 240 V
8 - 24 V
9 - 24 V
10 - 0 V
11 - 0 V
12 - 240 V

EXAMPLE 2

1 - 0 V
2 - 120 V
3 - 0 V
4 - 0 V
5 - 0 V
6 - 24 V
7 - 24 V
8 - 0 V
9 - 120 V
10 - 0 V

1 - Indoor fan motor - Plenum fan control
2 - Gas valve - Limit switch - Transformer - Fuse L1
3 - Plenum fan control

EXAMPLE 3

1 - 0v
2 - 0 V
3 - 0 V
4 - 0 V
5 - 24 V
6 - 0 V
7 - 24 V
8 - 24 V
9 - 0 V
10 - 0 V
11 - 24 V
12 - 24 V
13 - 240 V
14 - 0 V
15 - 240 V
16 - 0 V
17 - 24 V
18 - 120 V
19 - 0 V
20 - 0 V

1 - Plenum switch-fan relay
2 - Condenser fan motor
3 - Contactor-high pressure switch-low pressure switch
4 - Indoor fan motor-fan relay
5 - Thermostat contacts-subbase selector switch

EXAMPLE 4

1 - 0 V
2 - 0 V
3 - 0 V
4 - 10,000 V
5 - 120 V
6 - 0 V
7 - 0 V
8 - 0 V
9 - 120 V
10 - 0 V
11 - 0 V
12 - 0 V
13 - 0 V
14 - 0 V
15 - 0 V
16 - 24 V
17 - 24 V
18 - 24 V
19 - 24 V
20 - 0 V

1 - Fan motor
2 - Ignition transformer
3 - Burner motor
4 - Heating transformer - limit switch
5 - Cover cad cell, line 16

EXAMPLE 5

1 - 24 V
2 - 24 V
3 - 0 V
4 - 0 V
5 - 0 V
6 - 0 V
7 - 24 V
8 - 0 V
9 - 120 V
10 - 24 V

1 - Indoor fan motor - fan relay -thermostat contacts R to G
2 - Cooling contactor coil - thermostat contacts R to Y
3 - Indoor fan motor high speed winding - Fan relay - thermostatcontacts R to G

EXAMPLE 6

1 - 120 V
2 - 120 V
3 - 120 V
4 - 0 V
5 - 0 V
6 - 0 V
7 - 0 V
8 - 120 V
9 - 120 V
10 - 0 V

1 - Freezer light - freezer door switch
2 - Compressor - start relay
3 - Freezer door heater
4 - Fuses - service cord
5 - Thermostat

EXAMPLE 7

1 - 0 V, check switch position chart
2 - 0 V, check switch position chart
3 - 0 V
4 - 120 V
5 - 120 V
6 - 120 V
7 - 0 V
8 - 0 V, check switch position chart
9 - 0 V, check switch position chart
10 - 120 V

1 - Motor low speed winding - selector switch
2 - Compressor - start relay - thermostat - selector switch
3 - Selector switch - service cord
4 - Selector switch - fan motor
5 - Compressor - selector switch

EXAMPLE 8

1 - 0 V
2 - 0 V
3 - 0 V
4 - 0 V
5 - 0 V
6 - 240 V
7 - 0 V
8 - 240 V
9 - 240 V
10 - 0 V

1 - Contactor coil, line 7 -Time delay contacts, line 7 - Time delay heater, line 8
2 - Pump motor
3 - Indoor fan motor - Fan relay - Thermostat contact R to G
4 - Compressor - High-pressure switch -Contactor - Time delay relay
5 - High-pressure switch

EXAMPLE 9

1 - 240 V
2 - 240 V
3 - 0 V
4 - 240 V
5 - 240 V
6 - 24 V
7 - 0 V
8 - 24 V
9 - 24 V
10 - Below 24 V, voltage drop

1 - Condenser fan motor
2 - Compressor - line fuse L3
3 - High pressure switch -internal protector, OL1 - OL2 - OL3
4 - NO
YES
5 - YES
YES

EXAMPLE 10

1 - 240 V
2 - 0 V
3 - 0 V
4 - 24 V
5 - 24 V
6 - 0 V
7 - 24 V
8 - 0 V
9 - 0 V
10 - 0 V

1 - 2HTR Limit - Fuses - line 8
2 - Relay 2TR
3 - Fan motor - Fan relay
4 - Contacts 3TR-3, line 2
5 - NO
Contacts 2TR-2 must close to energize time delay relay coil 3TR, to close contacts 3TR-1 to energize 3HTR

Section 6 Solutions

RUUD

1. L1 to L2 on the ignition control board	115 V
2. Draft Motor IDM (IDM)	115 V
3. Limit ALC	0 V
4. Contacts K2 —||—	0 V
5. Contacts K1 NO —||— and K2 NC —|/|—	115V/0V
6. Transformer primary 115 FU	115 V
7. Contacts K3 —||—	0 V
8. Gas valve contacts? [GV]	24 V
9. Blocked drain switch BDS	0 V
10. Blower motor low speed to R	115 V

Problem 1

STEP 1. Has been completed.
STEP 2. Check for 115 V from L1 to Neutral.
STEP 3. Test fuse at fused disconnect.
STEP 4. Induced draft motor.
STEP 5. ☐ Check voltage at the draft motor terminals If 115 V ispresent, replace draft motor.
☐ Check that normally-open contacts K3 are closed
☐ If K3 is open, check inducer control for 24 V. If 24 V is present, replace inducer control.
☐ If no 24 V, check NPC inlet normally-closed contacts.

Problem 2

STEP 1. Has been completed.
STEP 2. Check for 115 V from L1 to Neutral.
STEP 3. Test fuse at fused disconnect.
STEP 4. Igniter element.
STEP 5. ☐ Check for 115 V at the ignition Control terminals and IGN and IGN.
☐ If voltage is present replace hot surface element.
☐ If no voltage is read, replace ignition control.

Problem 3

STEP 1. Has been completed.
STEP 2. Check for 115 V from L1 to Neutral.
STEP 3. Test fuse at fused disconnect.
STEP 4. Gas valve
STEP 5. ☐ Check voltage at terminals of gas valve.
☐ If 24 V is present, replace gas valve.
☐ If no voltage is read, check for 24 V at terminals, valve and GND, on the HS780 ignition control.
☐ If no voltage is found, check Interlock door switch BDS, NPC inlet, for open contacts.

APRILAIRE

1.	Furnace blower motor	115 V
2.	Humidifier fan motor	115 V
3.	Water Valve solenoid coil	115 V
4.	Electromagnetic relay coil	24 V
5.	Secondary of transformer	24 V
6.	Humidistat contacts	0 V
7.	Relay contacts	0 V

Problem 2.

STEP 1. Has been completed.

STEPS 2 and 3. Blower motor is running, voltage and fuses are all right.

STEP 4. Water valve and fan motor will not operate.

STEP 5. ☐ Check the transformer secondary. If you do not read 24 V, the transformer is defective.

☐ Check the relay coil. If 24 V is read, the relay is defective.

☐ Check the relay contacts. If 115 V is read, the relay is defective.

☐ Check the humidistat. If 24 V is read, the humidistat is defective.

☐ Check the water valve and the fan motor, although it is unlikely that they would fail together. If you read 115 V, the units are defective.

FRIGIDAIRE

COOL

1.	Freezer mullion drier, Line 1	115 V
2.	Door drier, Line 6	115 V
3.	Center vertical drier, Line 7	115 V
4.	Defrost heaters, Line 7	115 V
5.	Cold control heater, Line 8	115 V
6.	Freezer fan, Line 9	115 V
7.	Condenser fan, Line 11	115 V
8.	Defrost timer motor, terminals 1 to 3, Line 13	115 V
9.	Common to run on the compressor, Line 13	115 V
10.	Cold control, Line 9	0 V
11.	Motor protector, Line 10	0 V

DEFROST

1.	Freezer mullion drier, Line 1	115 V
2.	Door drier, Line 6	115 V
3.	Center vertical drier, Line 7	0 V
4.	Defrost heaters, Line 7	115 V
5.	Cold control heater, Line 8	0 V
6.	Freezer fan, Line 9	0 V
7.	Condenser fan, Line 11	0 V
8.	Defrost timer motor, terminals 1 to 3, Line 13	115 V
9.	Common to run on the compressor, Line 13	0 V
10.	Cold control, Line 9	0 V
11.	Motor protector, Line 10	0 V

Problem 1.

STEP 1. Has been covered.

STEPS 2 and 3. Since the box cools, the voltage and fuses are all right.

STEP 4. Refrigerator will not defrost.

STEP 5. ☐ Check the voltage across the defrost heater. If voltage is read, and the box needs defrosting, the heater coil is defective.

☐ Check the defrost thermostat. If voltage is read, the thermostat is open.

☐ Check for 115 V between terminals 3 and 4 of the defrost timer. If voltage is read, the timer motor is defective.

If no voltage is read, there is a broken wire between terminal 3 and B or between terminal 1 and W.

Problem 2.

STEP 1. Has been completed.

STEPS 2 and 3. Partial operation indicates that the voltage and fuses are good.

STEP 4. Refrigerator will not cease defrosting.

STEPS 5. ☐ Check between terminals 3 and 4 of the defrost timer.

If no voltage is read, the switch is closed. If voltage is read, the system should not be in defrost.

☐ Check the timer motor between terminals 1 and 3. If voltage is read, the motor is defective.

If no voltage is read, there is a broken wire... usually from terminal 1 to W.

Problem 3.

STEP 1. Has. been completed.

STEPS 2 and 3. The fans are working, so the voltage and fuses must be good.

STEP 4. Compressor will not start. There is no hum.

STEPS 5. The schematic shows that the compressor and both fans are switched on by the cold control.

☐ Check for voltage from the compressor common terminal to W. If voltage is read, there is a broken wire between these two points.

☐ Check the motor protector. If voltage is read, the protector is open.

If no voltage is read, wait for the internal overload to reset.

☐ Check the compressor relay between terminals 2 and 4. If voltage is read, the relay is defective. Otherwise, there is a broken wire between the compressor and its relay.

HEIL

1. Ⓡ to Ⓒ on the compressor 230 V
2. RUN to common on the condenser fan motor 230 V
3. Ⓛ to Ⓣ on the contactor 0 V
4. Ⓡ to Ⓨ on the thermostat 0 V
5. Ⓒ to Ⓨ on the condensing unit contactor coil 24 V
6. Primary of transformer 115 V

Problem 1.

STEP 1. Has been completed.

STEPS 2 and 3. Compressor is running, voltage and fuses are all right.

STEP 4. Condenser fan is not running.

STEP 5. ☐ Check terminals RUN to common on the condenser motor. If voltage is read, the internal protector is open or the run capacitor is defective or the motor is defective.

☐ Check the solid state, variable speed fan control by checking for voltage between the common terminal of the fan motor and terminal T of the contactor. If you read 230 V, the speed controller is open or there is a broken wire between the two points.

☐ Check the wiring from RUN on the condenser motor to the other terminal T on the contactor. If you read 230 V, the wire between the two points is broken.

Problem 2.

STEP 1. Has been completed.

STEPS 2 and 3. The wiring diagram shows that there are two field-wired power supplies: 115 V for the fan and 230 V for the compressor and condenser motors. Therefore, check the voltage and fuses for the 230-V supply.

STEP 4. The compressor and condenser motors are not running. Since the blower motor is operating, we know that there is 24 V available to both the fan center and the condensing unit contactor coil.

STEP 5. ☐ Check for voltage between terminals R and C on both the compressor and condenser motors. The odds are against finding both motors defective at the same time, but a 230 V reading will prove that is the case.

☐ Check for voltage between both sets of terminals L and T on the motor contactor.

If you read 230 V across one set of contacts, the contactor is defective. If you read 230 V over both sets of contacts, the contactor is open.

☐ Check the condensing unit contactor coil in the 24-V circuit. If you read 24 V, the coil is defective.

☐ Check for voltage between R and Y on the thermostat. If 24 V is read, the thermostat is defective.

☐ Check the wiring from Y on the thermostat to Y on the contactor coil. If 24 V is read, the wire in broken.

Problem 3.

Step 1. Has been completed.

Steps 2 and 3. Check the voltage and fuses for both the 115 V and 230 V supplies, since none of the motors are running.

Step 4. All the motors are inoperative.

Step 5. ☐ Check the 24 V transformer first, since it supplies the control circuit. If you read 115 V across the primary and no voltage across the 24 V secondary, the transformer is defective.

☐ Check across terminals R to Y and R to G in the thermostat. If you read 24 V, the thermostat is defective.

☐ Check from terminal R on the thermostat to the transformer. If you read 24 V, there is a broken wire between the two points.

SCOTSMAN

1. Master switch	0 V
2. Bin thermostat	0 V
3. Low pressure control	0 V
4. Spout switch	0 V
5. Fan control (motor at low speed)	220 V
6. Fan motor (motor at high speed)	220 V
7. Auger motor	220 V
8. Contactor coil	220 V

Problem 1.

STEP 1. Has been completed.

STEPS 2 and 3. The compressor is running, so there is no need to check voltage and fuses.

STEP 4. Fan motor is not running.

STEP 5. ☐ If there is 220 V at the fan motor terminals, the fan is defective.

☐ Check the fan resistor. If 220 V is read, the resistor is open.

☐ If there is no voltage at any of these points, look for a broken wire.

Problem 2.

STEP 1. Has been completed.

STEPS 2 and 3. Check the voltage and fuses. This is a 3-phase compressor. If the L_2 phase is out, the auger motor can still operate.

STEP 4. The compressor and condenser fan motor are inoperative.

STEP 5. ☐ Check the 3-phase contactor. If voltage is read across any of the contacts, L_1 to T_1, L_2 to T_2, L_3 to T_3, the contacts are defective.

If there is no voltage at any of the contacts, the contacts are open. With the contacts open, voltage can be read from L_1 to L_2 to L_3, but not from L_1 to T_1, etc.

☐ Check the contactor coil. If voltage is read, the coil is defective.

☐ Check the spout switch. If voltage is read, the switch is open.

☐ Check the voltage at the normally-closed overload at the auger drive.

If no voltage is present, there is a broken wire between the overload switch and the contactor coil.

Problem 3.

STEP 1. Has been completed.

STEPS 2 and 3. If nothing is running, first check the voltage and fuses on all three lines.

STEP 4. Nothing is operating.

STEP 5. ☐ The schematic shows three switches in series, each one of which could render the system inoperative. They are the master switch, the bin thermostat, and the low-pressure control. Check each of these switches. If voltage is read, the switch is open.

☐ If two or more switches are open, voltage can only be read from L_1 on the contactor to the auger motor terminal.

ARKLA

1. Relay contacts	0 V
2. Time-delay contacts	0 V
3. Sail switch	0 V
4. Chilled water switch	0 V
5. High-temperature switch	0 V
6. Gas valve coil	24 V
7. Time-delay relay coil	24 V
8. Relay coil	24 V
9. Pump motor	230 V
10. Run to common on fan motor	230 V

Problem 1.

STEP 1. Has been completed.

STEP 2. Check for 24 V at the transformer secondary.

STEP 3. If there is no 24 V, check the fuses to the transformer primary.

STEP 4. The gas valve will not open.

STEP 5. ☐ The gas valve has a 24-V solenoid coil. Check the coil. If 24 V is present, the coil is defective.

If there is no voltage, check the mechanical linkage within the valve body. It may be defective although the coil is good.

☐ Both the sail switch and the chilled water switch are in series with the gas valve. If 24 V is read across either switch, that switch is open.

If the sail switch is open, the fan motor is not running.

If the chilled water switch is open, it should close if the water temperature is 42 °F.

☐ Check the thermostat. If 24 V is read across terminals R to Y, the thermostat is defective.

☐ Check the high-temperature switch. If 24 V is present, the switch is open.

If no voltage is available, a wire is broken.

Problem 2.

STEP 1. Has been completed.

STEP 2. Check for 230 V between L_1 and L_2.

STEP 3. If there is no voltage, check the fuses.

STEP 4. Fan and pump motor will not start.

STEP 5. ☐ Check the voltage at the terminals of each motor. If 230 V is read at each motor, they are both defective.

If 230 V is read at only one motor, check for a broken wire.

☐ Check the relay contacts. If there is no voltage, a wire is broken.

If 230 V is read, the contacts are open.

☐ Check the relay coil in the 24-V circuit. If 24 V is read across the relay coil, the coil is defective.

☐ Check the time-delay switch. If there is no voltage, look for a broken wire to the transformer secondary.

If 24 V is read, the contacts are open.

☐ Check voltage to the relay heater. If 24 V is read, wait approximately one minute for the contacts to close. If they do not close, the TD relay is defective.

☐ Check the thermostat. If 24 V is read across R to Y, the thermostat is defective.

☐ Check the high-temperature switch. If 24 V is read, the switch is open.

If there is no voltage, a wire is broken.

YORK

COOLING

	Voltage
1. Coil -ww- CCH (Line 2)	230 V
2. Coil (M) (Line 14)	24 V
3. Coil (6R) (Line 13)	24 V
4. Contacts -||- M (Lines 3, 4, 7)	0 V
5. Contacts -|/|- HPC (Line 14)	0 V
6. Contacts 1OL, -|/|- 2OL, -|/|- 3OL (Line 14)	0 V
7. Contacts -||- 6R (Line 9)	0 V
8. Compressor Terminals 1 to 2, 2 to 3, 1 to 3	230 V
9. Condenser Fan Motor Terminals T1 to T2	230 V
10. Blower Motor Terminals	230 V

Problem 1.

STEP 1. Has been completed.

STEPS 2 and 3. Compressor is running, voltage and fuses are all right.

STEP 4. Blower motor is not running.

STEP 5. ☐ Check the voltage at the blower motor terminals. If 230 V is read, the motor is defective.

☐ Check the 6R relay contacts. If 230 V is read, the contacts are open.

If no voltage is available at the relay, there is a broken wire between L_1 and L_3.

☐ Check coil 6R in the 24-V circuit. If 24 V is read, the coil is defective.

☐ Check the thermostat. If 24 V is read from V to F, the thermostat is defective.

If no voltage is available at the thermostat, the wire from terminal F on the thermostat to terminal F on the package unit is broken.

Problem 2.

STEP 1. Has been completed.

STEPS 2 and 3. The blower motor will operate if fuse L_2 is open. The compressor and the condenser will not. Check voltage and fuses.

STEP 4. Compressor and condenser fan are not running.

STEP 5. ☐ Check the compressor and condenser fan motor terminals, even though it is improbable that both both motors are simultaneously defective. However, if you read 230 V, the motor is defective or the internal overloads are open.

☐ Check from the load side of contact M on Line 3 to the load side of contacts M on Lines 4 and 7.

If 230 V is read, the contacts are closed and one or more wires on the load side of the contacts is broken.

If no voltage is read, the contacts are open.

☐ Check coil M in the 24-V circuit. If 24 V is read, the coil is defective.

☐ Check the safety switches: 1OL, 2OL, 3OL, HPC. If 24 V is read across any switch, that switch is open.

☐ Check from terminal C to terminal T on Line 14. If 24 V is read, there is more than one safety switch open.

☐ Check the thermostat. If 24 V is read from terminals V to C, the thermostat is defective.

If no voltage is present, the wire from C in the thermostat to C in the air conditioner is broken.

Problem 3.

STEP 1. Has been completed.

STEPS 2 and 3. If the condenser fan and the blower motor are running, all three fuses are good.

STEP 4. Compressor is not running.

STEP 5. ☐ Check for voltage at the compressor terminals. If 230 V is read, the compressor is defective.

☐ Check from T1 to T3. If no voltage is read, contacts M on Line 7 are open and the contactor is defective, even though the other two contacts M are supplying power to the condenser fan.

If voltage is read, there is a broken wire.

☐ The overloads are apparently closed. If one of them were open, the condenser fan would not be operating.

HEATING

1. The HR coils (Lines 25A, 25, 24, 23, 22, 21)	20 vdc
2. Coil (3R) (Line 19)	20 vdc
3. Heater element 3H (Line 26)	230 vac
4. Heater element 5H (Line 28)	230 vac
5. Contacts 2TLS (Line 17)	0 V
6. Contacts 1TLS (Line 18)	0 V
7. Contacts 2HR (Line 26)	0 V
8. Contacts 6HR (Line 28)	0 V
9. Contacts 3HR (Line 8)	0 V
10. Blower Motor Terminals (Lines 8, 9)	230 vac

Problem 1.

STEP 1. Has been completed.

STEPS 2 and 3. Each bank of heaters is fused separately. The fuses must be checked.

STEP 4. Elements 4H, 5H, and 6H do not heat.

STEP 5. ☐ Check each heater. If 230 V is read, the heater is defective.

☐ Check fuses 4FL, 5FL, 6FL. If 230 V is read, the fuse is open.

☐ Check contacts 4HR, 5HR, 6HR. If 230 V is read, the contacts are open.

☐ Check relay coil 4HR on Line 23. If 20 vdc is read, the coil is defective. (Relays 5HR and 6HR need not be checked since they will not energize until 4HR is energized.)

☐ Check the time delay relay 3TR on Line 18. This relay must be energized to close a circuit to relay 4HR. If 20 vdc is read, the relay is defective.

If no voltage is available, a wire to the coil is open.

Problem 2.

STEP 1. Has been completed.

STEPS 2 and 3. Check the fuses. If L_2 is open, heater 1H will still operate.

STEP 4. Elements 2H and 3H will not operate.

STEP 5. ☐ Check heaters 2H and 3H. If 230 V is read, the heaters are defective.

☐ Check fusible links 2FL and 3FL. If 230 V is read, the fuses are open.

☐ If no voltage is available to the heaters, a wire is open from L_2.

☐ Check the 1TR relay. If 20 vdc is read, the relay is defective.

☐ Check the 2HR relay. If 20 vdc is read, the relay is defective.

☐ The succeeding relays should be all right, since they sequence from those that precede.

Problem 3.

STEP 1. Has been completed.

STEPS 2 and 3. Check voltage and fuses. The blower is not in the heater circuit.

STEP 4. Blower motor is not running.

STEP 5. ☐ Check the blower motor. If 230 vac is read, the blower is defective.

☐ Check the 3R contacts, Line 8. If 230 V is read, the contacts are open.

If no voltage is available, a field wire is open.

☐ Check the 3R relay coil, Line 19. If 20 vdc is read, the coil is defective.

☐ Check from terminal F to F2, Line 19. If voltage is read, a wire is open to coil 3R.

If no voltage is available, a wire is open inside the sequencer.

FEDDERS

COOLING

1. Coil CC1	24 V
2. Coil CC2	24 V
3. Coil	24 V
4. Coil RC RCH	24 V
5. Coil HC1	230 V
6. Contacts RC	0 V
7. Contacts RCH	0 V
8. Contacts CC1	0 V
9. Contacts CC2	0 V
10. Motor MCF	230 V
11. Motor MCH	230 V

Problem 1.

STEP 1. Has been completed.

STEPS 2 and 3. One compressor is running, voltage and fuses are all right.

STEP 4. Second stage compressor is not running.

STEP 5. □ Check compressor terminals R to C. If 230 V is read, the compressor is defective or the internal protector is open.

□ Check the load side between contacts CC2 and CC2. If 230 V is read, there is an open wire between the compressor and the contactor.

□ Check the line side of contacts CC2 and CC2. If 230 V is read, the contacts are open.

If no voltage is available, there is a line open to the contactor from L_1 or L_2.

□ Check coil CC2 in the 24 V circuit. If 24 V is read, the coil is defective.

If no voltage is available, there is a wire open from terminal CC2 to the coil.

Problem 2.

STEP 1. Has been completed.

STEPS 2 and 3. One compressor is running, voltage and fuses are all right.

STEP 4. First stage compressor is not running.

STEP 5. □ The schematic shows that the condenser fan motor is wired parallel to the first stage compressor. Therefore, the contacts are all right.

□ Check between terminals R and C on compressor 1. If 230 V is read, the compressor is defective.

If no voltage is available, a wire is open between the contacts and the compressor.

Problem 3.

STEP 1. Has been completed.

STEPS 2 and 3. Compressor is running, voltage and fuses are all right.

STEP 4. Blower motor is not running.

STEP 5. ☐ Check for voltage at blower motor terminals MCH. If 230 V is read, the motor is defective.

☐ Check across relay contacts RCH. If 230 V is read, the contacts are open.

If no voltage is available, a wire is open to L_1 or L_2.

☐ Check across coil RCH. If 24 V is read, the coil is defective.

☐ Check across contacts RC. If 24 V is read, the contacts are open.

If no voltage is available, a wire is open to terminal V.

☐ Check coil RC. If 24 V is read, the coil is defective.

☐ Check across terminals V to F on the terminal block. If 24 V is read, the thermostat is open.

If no voltage is read, a wire is open from V to coil RC.

HEATING

1.	Coil RCB	24 V
2.	Coil RCH	24 V
3.	Coil GV	24 V
4.	Coil GC	0 V
5.	Switch SGP	0 V
6.	Switch SS	0 V
7.	Switch SF	0 V
8.	Contacts RCB	0 V
9.	Contacts RCH	0 V
10.	Motor MCB	230 V
11.	Motor MCH	230 V

Problem 1.

STEP 1. Has been completed.

STEPS 2 and 3. Check line voltage fuses and 2.5-amp fuse at transformer secondary. Check for 24 V at transformer secondary.

STEP 4. Burner will not ignite.

STEP 5. ☐ Check the gas valve. If 24 V is read, the gas valve is defective.

☐ Check across the safety switches: LS, ALS, SS. If 24 V is read across any switch, that switch is open.

If LS or ALS is open, give them time to reset.

☐ If SS is open, check the MCB motor terminals. If 230 V is read, the motor is defective

☐ Check across contacts RCB. If 230 V is read, the contacts are open

☐ Check coil RCB. If 24 V is read, the coil is open.

☐ Check switch SGP. If 24 V is read, the switch is open.

If no voltage is available, a wire is open from H to the switch.

Problem 2.

STEP 1. Has been completed.

STEPS 2 and 3. With partial operation, voltage and fuses are all right.

STEP 4. Blower will not operate while heating.

STEP 5. ☐ The schematic shows that the blower motor is controlled by a fan switch in the 24-V circuit.

☐ Check the SF switch. If 24 V is read, the switch is open.

If no voltage is available, there is a wire open from terminal V to the switch.

Problem 3.

STEP 1. Has been completed.

STEPS 2 and 3. Voltage and fuses are all right.

STEP 4. Pilot flame will not ignite.

STEP 5. ☐ Check the glow coil. If 2.5 V is read, the coil is defective.

☐ Check across relay contacts RAP. If voltage is read, the contacts are open.

If no voltage is available, a wire is open from the transformer to the RAP relay or the 2.5-V transformer secondary is open.

WILLIAMSON

1. Motor relay coil 24 V
2. Flame relay coil 24 V
3. Fan motor 115 V
4. Burner motor 115 V
5. Fan switch 0 V
6. Motor relay contacts 0 V
7. Flame relay contacts 24 V
8. Filter flag 24 V
9. Humidistat (open) 24 V

Problem 1.

STEP 1. Has been completed.

STEPS 2 and 3. With partial operation, voltage and fuses are all right.

STEP 4. Fan motor will not run during cooling cycle.

STEP 5. ☐ Check the fan motor. If 115 V is read at the terminals, the motor is defective.

☐ Check the fan relay contacts. If 115 V is read, the contacts are open.

If no voltage is present, there is a broken wire.

☐ Check the fan relay coil in the 24-V circuit. If 24 V is read, the coil is defective.

☐ Check the thermostat across terminals R to G. If 24 V is read, the thermostat is defective.

If 24 V is not read, there is a broken wire, usually from G to the fan relay coil.

Problem 2.

STEP 1. Has been completed.

STEPS 2 and 3. Burner motor runs, voltage and fuses are all right.

STEP 4. Burner motor and ignition cease after 45 seconds.

STEP 5. ☐ If the flame detector does not see a flame in 45 seconds, the burner and ignition will cut out in 45 seconds. First, check the oil supply.

☐ Check the limit switch. If 115 V is read, the limit is open and may be defective.

☐ Check the flame detector following the manufacturer's recommendations.

GE

COOLING

1.	Coil MS	24 V
2.	Coil F	24 V
3.	Coil	24 V
4.	Switch HPCO	0 V
5.	Switch F-1	0 V
6.	Switch DFC	0 V
7.	I.D. Fan Motor A to M	230 V
8.	CPR Motor C to R	230 V
9.	O.D. Fan Motor HI to WHITE	230 V

Problem 1.

STEP 1. Has been completed.

STEPS 2 and 3. Compressor is running, voltage and fuses are all right.

STEP 4. Hot rather than cool air during summer operation.

STEP 5. □ The schematic shows that the switch-over valve is energized with the first stage of cooling, the second stage starts the compressor.

□ Check for voltage across the switch-over coil (SC). If 24 V is read, the coil is defective or the switch-over valve is stuck in the heating position.

□ Check the low-voltage terminal board from R to O. If voltage is read, the thermostat is open.

□ If no voltage is read from between the thermostat and the switch-over coil, there is a broken wire.

Problem 2.

STEP 1. Has been completed.

STEPS 2 and 3. Indoor fan is running, voltage and fuses are all right.

STEP 4. Compressor and outdoor fan are not running.

STEP 5. □ Check from C to R on the compressor motor and from Red and Black to White on the outdoor fan motor. If voltage is read, the motors are defective or the internal protectors are open.

□ Check for voltage across contacts MS-1A and MS-1B. If 230 V is read, the contacts are open. If no voltage is read, look for a broken wire.

□ Check coil MS in the 24-V circuit. If 24 V is read, the coil is defective.

□ Check the high pressure cutout. If 24 V is read, this safety device is open.

□ On the low-voltage terminal board, check from R to Y. If 24 V is read, the thermostat is open.

If voltage is read from Y to the MS coil, look for broken wires between these two points.

Problem 3.

STEP 1. Has been completed.

STEPS 2 and 3. Compressor is running, voltage and fuses are all right.

STEP 4. Outdoor fan is not running. The indoor fan is running and the compressor is cycling on the high-pressure cutout.

STEP 5. ☐ Check for voltage between Black and Red to White at the outdoor fan motor. If 230 V is read the motor is defective.

☐ Check contacts MS-1B. If 230 V is read, the contactor is defective.

If voltage is not present, look for a broken wire to the motor.

HEATING

1. Coil MS	24 V
2. Coil F	24 V
3. Coil D	0 V
4. Contact DFC to HI	0 V
5. Contacts F-1	0 V
6. Contacts F-2	0 V
7. Contacts MS-1A	0 V
8. Contacts MS-1B	0 V

Problem 1.

STEP 1. Has been completed.

STEPS 2 and 3. Outdoor fan is running, voltage and fuses are all right.

STEP 4. Compressor is not running. Fans are running.

STEP 5. ☐ Check for voltage at compressor terminals R and C. If 230 V is read, the compressor is defective or the internal protector is open .

☐ Check contacts MS-1A. If 230 V is read, the contacts are open and the contactor is defective.

☐ If no voltage is present, a wire is broken between the motor and the contacts.

Problem 2.

STEP 1. Has been completed.

STEPS 2 and 3. Check voltages, including the 24 V secondary, and the fuses.

STEP 4. Nothing is running.

STEP 5. ☐ Check across the compressor terminals R to C, the outdoor fan motor terminals Black and Red to White, indoor fan terminals Black to Black. If 230 V is read, the motor is defective.

If no voltage is present, there is a broken wire between the fuses and the motors.

☐ Check across contacts MS-1A, MS-1B, and F-1. If voltage is read, the contacts are open.

☐ Check the contactor coil MS and the fan coil F in the 24-V circuit. If 24 V is read, the coils are defective.

☐ Check the terminal board from terminal R to terminals Y and G. If 24 V is read, the thermostat is open.

If no voltage is present, a wire is open from R to the transformer secondary.

Problem 3.

STEP 1. Has been completed

STEPS 2 and 3. Each duct heater has a separate power supply. Check both voltage and fuses.

STEP 4. One duct heater will not operate. Contacts TSH-2 are closed. All motors are operating.

STEP 5. ☐ Check voltage at the duct heater terminals. If 230 V is read, the heater is defective.

☐ Check for voltage across the outdoor thermostat ODT contacts. If 24 V is read, the thermostat is open.

If no voltage is present, a wire is open between the ODT and the heater control relay.

Author Stanley Aglow has had more than 15 years of experience as an installation and service supervisor and writes a regular electrical service column for the *Air Conditioning, Heating and Refrigeration News.* Today, Mr. Aglow is the teacher and Director of HVAC/R Training at the Philadelphia Wireless Technical Institute.

Appendix

Example 1
Air Conditioning Compressor

L_1 L_2

Example 2
Gas-fired, Forced-air Furnace

L_1 N

Example 3
Gas-fired, Forced-air Furnace With Automatic Air Conditioning

L_1 L_2 N

Example 4
Oil-fired, Forced-air Furnace

L_1 N

Example 5
Oil-fired, Forced-air Furnace with Automatic Air Conditioning

L_1 L_2 N

Example 6
Two-Door Refrigerator / Freezer

L_1 N

Example 7
Window Air Conditioner

L_1 N

Example 8
Package Air Conditioner with Water-Cooled Condenser

L_1 L_2

Example 9
Single Package
Air-Cooled Air Conditioner

L_1 L_3 L_2

EXAMPLE 10
ELECTRIC HOT-AIR FURNACE

L_1 L_2